G. Augenbroe

404 - 894 - 1686

Building Design Management

D0218730

C. Anglebroc
404-844-1682

Building Design Management

Colin Gray and Will Hughes
Department of Construction Management & Engineering,
The University of Reading, UK

OXFORD   AUCKLAND   BOSTON   JOHANNESBURG   MELBOURNE   NEW DELHI

Butterworth-Heinemann
Linacre House, Jordan Hill, Oxford OX2 8DP
225 Wildwood Avenue, Woburn, MA 01801-2041
A division of Reed Educational and Professional Publishing Ltd

⌐Q   A member of the Reed Elsevier plc group

First published 2001

© Colin Gray and Will Hughes 2001

All rights reserved. No part of this publication may be reproduced in
any material form (including photocopying or storing in any medium by
electronic means and whether or not transiently or incidentally to some
other use of this publication) without the written permission of the
copyright holder except in accordance with the provisions of the Copyright,
Designs and Patents Act 1988 or under the terms of a licence issued by the
Copyright Licensing Agency Ltd, 90 Tottenham Court Road, London,
England W1P 0LP. Applications for the copyright holder's written
permission to reproduce any part of this publication should be addressed
to the publishers

British Library Cataloguing in Publication Data
Gray, C. (Colin)
   Building design management
   1. Architecture – Management
   I. Title   II. Hughes, Will
   720.6'8

ISBN 0 7506 5070 2

Library of Congress Cataloging in Publication Data
A catalogue record for this book is available from the Library of Congress

ISBN 0 7506 5070 2

Composition by Genesis Typesetting, Laser Quay, Rochester, Kent
Printed and bound in Great Britain by MPG Books Ltd, Bodmin, Cornwall

PLANT A
TREE
BTCV
British Trust for
Conservation Volunteers

FOR EVERY TITLE THAT WE PUBLISH, BUTTERWORTH-HEINEMANN
WILL PAY FOR BTCV TO PLANT AND CARE FOR A TREE.

Contents

Part Two: Design Management Practice

Preface

Since the original version of this book was published as a handbook for designers and managers of and within design organizations, considerable development in this area has taken place. The first handbook achieved considerable success and wide dissemination and has helped to form the subject and achieve a recognition that this is an important area in modern construction projects, if not one of the most important. Considerable advances have been made, but there are still few examples of total success. This is why we have decided to update the original report and write this book.

The construction industry has moved on. Professional Construction Management has emerged in the UK as a major force for change in conventional practice, predominantly to manage the design process in a better way. Buildings are more complex and yet there is a major thrust to reduce cost, provide better quality and greater certainty of delivery. These all have an impact on the design process, in many cases making it more complex and less manageable.

> 'I find myself intolerant of management books that seek to prescribe exactly "how it should be done". My own experience shows that there are many different ways of achieving one's aims and many different ways of leading an industrial company...
>
> ...Each one of us has to develop our own style, and our own approach, using such skills and personal qualities as we have inherited. What each of us does over a long period of trial and error is to acquire a set of tools with which we are comfortable and which we can apply in different ways to the myriad problems which we need to solve.'

> (Harvey-Jones 1988)

This book examines the process of designing construction projects in order to help to understand and manage the process in a better way. There is no single prescription that will suit every project and what we have attempted to do is to provide illustrations and guidance based on the current level of understanding of the design process. As Sir John Harvey-Jones rightly says, it is for the individual to apply the set of tools to the particular project.

Our main objective has been to help to improve the efficiency of the design process and its integration with the construction process. It is the responsibility of everyone concerned with a project to be aware of the issues, managerial needs, and required practice. We hope, therefore, that this book will appeal to a wide readership throughout the construction industry.

Since the original book was published there has been a growing recognition of the importance of this subject. There have also emerged a number of design managers. This was not our intention because we viewed the task of managing the design as the responsibility of everyone on the project. What was missing was an understanding of the design process and the way information flows between the people and organizations before construction starts. This is still our firm view, although we do recognize that the design process is now so complex that when everyone is doing his or her own thing, the vital need to produce a co-ordinated set of information for construction may be missed. So from taking a general view we have now taken a much more specific approach which not only considers the process but also puts forward ideas and strategies for achieving best practice.

We must emphasize that this book is not about how to design, but how to manage. Also, passing references are made to legal requirements such as health and safety law, building control legislation and to planning law. Although these references are made in the context of UK legislation, they serve as reminders to readers in any jurisdiction that they need to take account of these social requirements at the appropriate points in the process. Detailed accounts of these different aspects of legislation are not provided in this book as they are seen as constraints within which design has to operate and they are dependent upon the operational jurisdiction of the project.

We are grateful for the encouragement and help which we have received from clients, consultants and specialists during the preparation of this book. We would also like to thank the publishers and the anonymous reviewers who provided comments on early drafts of the text.

C. Gray and W. Hughes

Introduction

The approach to the design and construction of new buildings, in particular commercial and industrial buildings, has changed rapidly over the last 10 years. This has been largely due to clients expecting better performance from both designers and contractors and their need to be certain of the final outcome of their projects. Clients have also sought greater control over their projects and an increased involvement in decision-making. During this period several significant changes have occurred.

- Construction projects have changed to place a greater emphasis on the management and co-ordination of specialist designers and works contractors.
- Specialization has allowed firms to achieve market prominence through greater knowledge and competence in providing particular products or services, which has helped them achieve profitability and reduce risk.
- The traditional role of the architect has changed from project leader and manager to leader of the design team. This has led to considerable ambiguity about the leadership of the project as a whole, particularly where roles and responsibilities have not been clearly defined from the outset.

The ways in which contributions to the design process have changed are compared in Table 1.

Design is a complex process that continues to grow in complexity because of the dramatic increase in specialist knowledge. There are now many contributors to the design of a project from a wide variety of organizations. This gives rise to design processes that consist of a continual exchange and refinement of information and knowledge. Even the most experienced design teams can fail to manage this complex process and supply information at the wrong time and of the wrong quality to members of the production team. Today, a very large proportion of a building's components are made in factories and assembled on site. This is completely different from the handcrafted, site-based methods on which architectural practice was founded. This fundamental change has caused the designers to specify, or draw, every aspect of the

Table 1. The significant changes in the role of the architect and other professional designers

Yesterday	Today
Architects and engineers dominate the market, offering a professional service	New designers are emerging and the established design professions are facing redefinition
Architects hold the dominant position of authority in the design process	Architects are losing position and authority within the design team to managers, specialist designers, services and other engineers
Design and creativity are the dominant features of architectural education	Design education does not emphasize the ascendant skills, i.e., has not changed with the need
	Designers are working in settings that are not of their own design or control
Professionalism is based on narrow specialism	Professional designers are required to become generalists with less control over details, which are now dealt with by experts in a wide variety of fields
Architects are the natural leaders of the process	There is multiple control of the whole design and construction process
Fee agreements are simple and loose	Fee agreements are complex and restrictive
Designers determine the client's 'real' problem	Customer dominance by expert clients
High design quality	Value through design quality/cost/time trade-offs
Professionals are relied upon to deliver a competent service	Others manage designers
The professional designer has an overall responsibility for the management of the whole process	A wide variety of sophisticated procurement techniques attempt to integrate design and construction

project to a level of detail which removes all ambiguities in design intent from the manufacturing and site assembly processes, which in turn has led to an apparently insatiable demand for drawn information.

In other words, building design has now become an integral part of a complex industrial process and there is a need to identify the management task and manage it well. This book is designed to help designers and engineers who have to manage their own design process and design managers who have the task of managing a design team or group of designers. In all situations careful integration of the design process with the procurement and construction of the project is of paramount importance. However, for the efficient provision of design information, it may be necessary to subordinate the concerns of individual organizations to the demands of the project as a whole. Good design management helps designers to focus on the project needs and to be aware of the controlling activities. In practice, each member of the project team, which may consist of consultants, specialist trade contractors and project managers, will carry out a design management function which should be controlled by a manager from within each organization. Even so, there has to be a single point of responsibility on a project for the final delivery of approved production information for construction. The responsibility for this should be decided at the outset and the appropriate authority established.

Two issues should always be addressed: the provision of accurate, fully co-ordinated and complete information, and the timely provision of the information. The first issue is the responsibility of the lead designer and the second is that of management. Ideally they should be synonymous, but the complexity of both tasks now requires that the management of the total design process is identified as a fundamental need in modern projects and vested in individuals or organizations who are experienced in, and understand, the integrated process of design and construction.

While there are as many ways to design as there are designers, there is a consensus that there are some basic patterns. Form, function and fit are the key elements (Dumas and Mintzberg 1992). Form relates to style, function concerns engineering, and fit is the link between form and function. The success in any project is the management of the content within each area and the interface between them. Dumas and Mintzberg have proposed four management models.

- *Encompassed design: single function* – This is where the designers carry out the whole process in an integrated way. The organization manages all of the designers and does not have to manage the interfaces between different types of designer.
- *Decomposed design: isolated function* – The easiest way to manage the interfaces between designers is to decompose the design into clear components of function and form and then assign each to a group of designers. Linkages get taken care of on the drawing board before detailed design begins. This makes the designer's job easier, but its application is limited to circumstances where the parameters of design are well known and easily controlled, i.e., where little innovation or creativity is expected. This model works where mature products are applied in stable conditions.
- *Dominated design: leading function* – This approach attempts to supersede the problem of interfaces by supplanting them with a hierarchy. One group takes charge, to impose the design realization on the others. This 'over the wall' approach requires that the others have to conform to the needs of the dominant designer. Alternatively, a designer develops a vision that forces all others to integrate their work under this strong direction. Frustration and sub-optimal design often occur, as people are restricted in the contribution that they can make.
- *Co-operative design: interactive functions* – This model encourages interaction between the different contributors. Co-operative design is based on teamwork and reflects the ad hoc structure of most 'creative' organizations. This approach requires a number of mechanisms, teams, task forces, and integrating managers to promote mutual adjustment among experts, under conditions that are both dynamic and complex.

This final model is the one that is used in this book. While the other models are effective in specific circumstances the growing level of complexity makes them less appropriate.

The structure of the book

The book is divided into two parts. Part 1 reviews the design process and contains useful information for those new to the task of managing design and for those already familiar with the design process. Part 2 describes the detailed management techniques, many of which rely on a comprehensive understanding of the modern design process explored in Part 1.

Part 1 begins by dealing with the essential structure of the design process and considers the underlying method of a designer's approach to design problems. It continues by going on to consider how technology is applied to the normal variety of building projects and how it affects the complexity and scale of the design task. A description of the design and construction process is given, divided into the stages required for good management, together with the roles of client, designers and managers. We introduce the concept that the responsibility for managing the design process alters as each new stage is reached. We consider strategic issues and the complex subject of design team selection, including all inputs to the design, obtaining compatible agreements and the most appropriate organization to deliver the design information. Many projects, even quite small ones, use an architect supported by other professional consultants and so it is necessary to consider the whole question of team building and welding the diverse and separate contributors to the design into a coherent and collaborative team.

Part 2 describes the tools, processes and actions for the improved management of the design process as part of an integrated system, which gives a total quality approach to the management of design. This addresses the scale of the information requirements, and detailed checklists are provided from which individual design managers can develop their own analysis of the tasks to be done. The design, as it is being developed, should be evaluated to ensure that it produces value for money, simplicity of construction and long-term satisfaction in operation. Descriptions and checklists for the specialized techniques and processes of these evaluations are provided. Finally, the basic techniques of planning, monitoring and control appropriate for managing the specific requirements of the design process are considered and new methods are proposed.

Part One
Design management theory

1

Management and organization in design and construction

There are many ideas and concepts in the literature on organization and management that provide the context and background for thinking about how to manage work within a specific domain of activity. In considering how to manage the design process in construction it is useful to begin by reviewing some basic concepts from management thinking.

The idea of separating responsibility for managing from the responsibility for doing work is firmly entrenched in the management literature. This has been very effective in the management of industrial production processes, and can be traced back to Adam Smith and Charles Babbage, as shown by Hawk (1996). In construction, especially when using general contracting, the phenomenon is clearly visible in the separation of responsibility for designing from the responsibility for construction. It is given further focus because, historically at least, the designer has supervised the construction of the work. In practical terms, this has come to signify that design and management are two sides of the same coin. However, there is a great deal of evidence to support the idea that while it is important for designers to be the managers of the construction process, the management of the design process is something inherently different from the act of design. This is borne out in the RIBA Plan of Work, the new (draft) JCT Consultants' Agreement, and the Latham Report, all of which identify explicitly the two roles of designing and of managing.

Dumas and Mintzberg (1992) refer to an inherent danger embedded in the idea of design management. Although design is primarily undertaken by designers, decisions made by design managers may have a profound influence on what the designers do. This they call 'silent design development'.

In the UK, the division of responsibility for design from the responsibility for construction is highly institutionalized and underpins a wide range of professional and commercial activity. This division was famously commented upon in the Banwell Report (1964) and is so well entrenched that even apparently integrated construction management practices split the running of their projects between their design managers and their construction managers.

This book is about the management of the design process. Before looking in detail at the problem of design management, this chapter introduces a series of management concepts that are used throughout the book. A glossary of terms for easy reference can be found at the end of the book.

1.1 Organizational issues

1.1.1 Organization vs. management

The origin of the word 'organize' is related to the notion of organs in the human body. To organize something is to arrange the elements into a co-ordinated whole. This shows that complex things can only be understood when orderly structure is imposed upon them. In other words, dealing with complex issues often requires the whole to be split into pieces. Of course, great care is needed in dealing with human and social systems in order that this kind of disassembly does not destroy their essence. Indeed, there can be enormous resistance to excessively rational solutions to human and social problems, as Morgan (1986, p. 38) observed in his comment that the mechanistic approach to organization tends to limit rather than mobilize the development of human capacities.

The definition of 'manage' is to conduct things and people in order to achieve some end. It comes from the Latin word *manus*, meaning hand. Manage has also come to mean accomplish, usually successfully, but it sometimes has the meaning of barely coping, depending upon the context. Management involves co-ordination, motivation, leadership and many aspects of getting things done through other people.

1.1.2 Contingency theories

Contingency theories are based upon system theories in that the organization is identified as an open system defined by a boundary outside of which is the 'environment'. Within the boundary is a set of related parts that must interact in some useful way to produce some kind of output. The set of parts is what defines the structure of the organization. The output from the organization is destined for the environment where it will have an impact of some kind. The fact that the environment is the source of inputs for the organization and surrounds its activities and absorbs any outputs, means that an understanding of the relationship between the environment and the organization is essential in understanding how best to organize an activity.

In construction project terms, the participants in the process form a temporary organizational structure for the purposes of the project. The links between participants are brought into sharp relief because they cross boundaries between firms. The project organizational structure is oriented towards effecting some sort of change beyond the project itself. For this reason, it is vital that the project is organized in the context of its environment.

1.2 Environment

The environment comprises everything that is not a part of the system under consideration. In distinguishing what is outside from what is inside, one simply needs to imagine a line enclosing the system under consideration. This line is drawn wherever it seems most useful for the particular analysis. In organizations, it is not too difficult to imagine that outside the line are things over which managers have little control and/or influence, and inside it are things over which managers have more control and/or influence. Deciding where this boundary is and what to do about it, are serious tasks for management at different levels. At the most senior level, the definition of the boundary between the project and its environment helps to define the purpose of the project organization and its place in regard to the rest of the world. This is a question of policy. At the strategic level, the managers of a project need to regulate flows across the boundary (in the open systems view, the critical element is the modelling of boundary transactions). At the tactical level, work undertaken within the project needs to be done within the context of how the project organization is seen in relation to the environment. Although these ideas have been developed in relation to the management of firms, research has shown that they are equally applicable to the management of projects (Hughes, 1989).

1.2.1 The environment influences the organization

The definition of the boundary of an organization takes account of the way in which the environment influences the project. The context of a project organization can be viewed from a number of perspectives; physical, social, cultural, economic and political. Each of these is important for different reasons. Understanding the way in which these things have an impact on the organization is critical in defining the role and purpose of the project. Similarly, understanding their impact on a project tells us much about how to define, organize, and manage the work.

1.2.2 The environment is influenced by the organization

The organization is a part of its environment. And, as the environmentalist movement would constantly remind us, all organizations have an impact on the (physical) environment. Just as an organization needs to develop a policy for ensuring that atmosphere, water supply and land are not polluted, so the organization needs to develop a view as to what impact is intended on its social, economic, cultural and political environment. This constitutes a loop. The organization undertakes work as a response to its environment and transforms inputs into outputs via some process so that the environment will be modified in some way. This means that any organization deliberately sets out to have an impact. This is the view of the project strategist – to decide the nature of the impact required. Ultimately this will underpin the development of a design philosophy for a project.

1.2.3 Scanning

Figure 1 depicts interchanges between the project and its environment. One critical feature is that while the environment can be seen in terms of inputs and outputs to and from the project, there is constant interaction between the project organization and its environment. The environment is dynamic, constantly changing and subjectively defined. Therefore, what might have seemed appropriate yesterday may no longer be appropriate tomorrow. This is one reason that so many management writers stress the importance of change and of organizational learning. Innovation is partly concerned simply with developing new ways to help an organization adapt to its changing environment. If the organization is contingent upon its environment, then it is clear that *there is no single best way to organize*. Moreover, if the key variable upon which appropriateness depends is itself in a state of flux, then *the best way to organize a particular project will change from time to time* (Mintzberg, 1991).

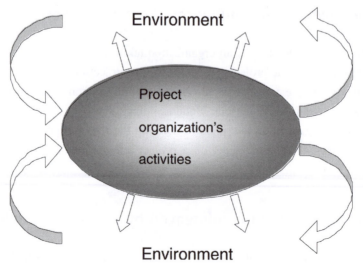

Fig. 1. An organization interacts with its environment.

1.2.4 Fit

The purpose of analysing the environment is to determine the most appropriate way of organizing the work. Goodness-of-fit is not a question of seeking an ideal or perfect solution but merely of ensuring that what is provided is the best reasonable fit under the circumstances (Dawson, 1996). Moreover, as the circumstances change, so the goodness-of-fit of the project organization will change.

1.3 Complexity

Complexity is common in construction projects (Bennett, 1992). However, it is not (usually) the result of technological complexity. The need for many different disciplines to come together during the design process is often compounded by the process of specialization, and the economic and professional pressures for each of these diverse disciplines to belong to a different firm. The same is true in construction: main contractors can no longer maintain a supply of work for their operatives. It is uneconomic to pay the wages of bricklayers, joiners, and all the other construction trades as it is not feasible for contractors to provide a continuous flow of work for these people. This is why subcontracting in construction is so widespread. So, for different reasons, participants at the design stage and at the construction stage, tend to be from different firms. This leads to a high degree of organizational complexity of the process and a tendency for complexity to continue to increase, as well as a lack of effective tools for measuring complexity (Gidado, 1996; Trinh and Sharif, 1996).

As will be seen later in this chapter, such complexity should not be avoided, but is a necessary part of a flexible and responsive industry. It is not the presence of complexity that is a problem, but the inability of project managers to deal with it (Southwell, 1997).

1.3.1 Technology

Dawson (1996) defines technology as the materials and processes used in transforming inputs into outputs. She adds into this definition the skills, knowledge and labour that an organization 'possesses'. Thus, technology has both hard and soft components, machinery and people, or in Dawson's terms, materials and operations on the one hand, and knowledge on the other. Technology may be described as a combination of the accumulated scientific knowledge, technical skills, implements, logical habits and material products of people, but it is always more than this, more than information, logic, and things. It is people themselves, undertaking their various activities in particular social and historical contexts, with particular interests and aims.

It is useful to note that the first industrial systems were developed by engineers, and it is they who were the heroes of the industrial revolution. The likes of Taylor, who developed the large machinery of production and had formal training as engineers, brought with them a particular world view which was mechanistic, objective and rational. As always, the production processes that were developed were ultimately dependent on the outcome of debates about what is available in terms of knowledge, expertise, ability to pay, materials, and energy. In addition, this was set into the context of what was seen as desirable, appropriate and advantageous at the time.

1.3.2 Professionalism and objectivity

Any discussion of judgement in terms of what seems desirable, appropriate and advantageous is, ultimately, subjective. But rational people typically feel that by exercising their professional judgement impartially, they are being objective. As Dawson (1996) said,

> 'this is particularly true for those who have had a fairly narrow education in science and engineering as they feel that scientific values and discoveries and their application merely reflect the inexorable and somehow neutral or independent march of progress.'

In fact, subjectivity prevails in such situations. Thus, what is seen as advantageous must be seen in the context of the observer. From the point of view of the professional in a construction project context, the key point is that professionals exercise discretion, impartially, but subjectively. If a decision could be taken purely on objective grounds then there would be no need for professional judgement and the task could be relegated to a technician or a computer.

1.3.3 Fragmentation and specialization

As work becomes more complex, so more diverse skills are needed to accomplish it (Lawrence and Lorsch, 1967b). The modern construction industry has its roots in an essentially craft-based process that consisted of only half a dozen different trades (Shand, 1954). The industrial revolution made more materials available, across a more dispersed area. New skills emerged to deal with these new materials. The pace of technological achievement has increased, and the modern construction project harnesses dozens, if not hundreds, of different skills during its design and fabrication. In general, the increasing speed of technological change looks set to continue (Handy, 1990).

A particular effect of specialization has been the relentless erosion of what used to be a single role in the management of construction projects, that of the architect/engineer

(A/E). Some of these changes have been brought about because of the demands of clients, some because of the demands of technological complexity and some because of institutional defensiveness. The progressive reduction in responsibility can be traced through the developments of general contracting, quantity surveying, town planning, accountancy, structural engineering, services engineering, project management, construction management, and contract adjudication. One way of illustrating the fragmentation of what was once a unified role is shown in Figure 2. This is intended to be indicative, rather than comprehensive.

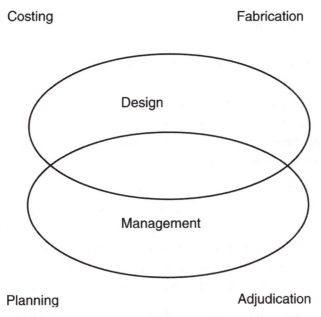

Fig. 2. Fragmentation of the traditional architectural role.

Financing and accountancy procedures in large organizations lead to ever more complex project planning phases before architects and engineers are appointed. The emergence of general contracting relieved Victorian architects and engineers of the need to manage the fabrication process, while leaving them with the responsibility for managing the design team. Similarly, the emergence of quantity surveying enabled clients to engage consultants who could help them maintain much closer control over expenditure than had been the case in the past. A more recent development has been the emergence of Adjudication (Housing Grants, Construction and Regeneration Act 1996), which challenges the impartial decision-making role that has traditionally been within the purview of the A/E. This role is assigned to another consultant appointed solely for the purpose of dealing with certain kinds of dispute. Obviously, other specialities can be added to this picture, but in general they usually fall into a more central position, as shown in Figure 2. This central region of design and management indicates the fact that although it may be theoretically possible to separate these two activities, in practice they are inextricably linked. It is interesting to note that a client who employs all of these specialists has little need of an A/E other than as a co-ordinating designer.

As a consequence of these divisions of responsibility, there is an increasing demand for managing and co-ordinating the diverse contributions to the project, an important issue for whoever is managing a construction project.

1.3.4 Differentiation

The concept of differentiation was introduced by Lawrence and Lorsch (1967a). They investigated the kinds of organization that were best suited to particular environments. Their two key variables were organizational structure and environment. Differentiation was defined as more than the mere division of labour or specialization. It also refers to differences in the attitudes and behaviour of the managers concerned. These they looked at in terms of (among other things):

- orientation towards certain goals (e.g. cost reduction is more important in Production than in Sales; or, more pertinent in construction, economy is more important in cost control than it is in structural design, where safety and stability are more important);
- time orientation (e.g. Sales and Production departments are likely to have a shorter term horizon than Research departments, and construction site planning is likely to have a shorter term than the development of a design brief).

Thompson (1967) developed a very specific list of types of differentiation: technology, territory, and time, to which he added two types of sentience as reinforcers of differentiation. Thus there are different perceptions of the same simple idea. The simple idea is that the people working on a particular thing differ in a number of respects. Lawrence and Lorsch's view focused on how people differ because of their own leanings and predilections, whereas Thompson's variables are more about external observations of how people differ. The key thing here is that these differences exist because they are needed.

The first thing in Thompson's list, technology, is the key variable picked up by Lawrence and Lorsch and by Woodward (1965). For Woodward, technology was the most important influence on the way that the most successful firms were organized. For Lawrence and Lorsch, technology was the starting point, but they recognized others. They identified that greater levels of differentiation required greater levels of integration. The amount of differentiation required was dependent on the complexity of the organization's environment, so simply eliminating differentiation is of no help.

1.3.5 Integration

Each additional member of a team increases the demand for integration. To ensure that the diverse contributions of the team members are mutually sympathetic, information needs to flow between all members of the team. Integration is concerned with unifying the diverse contributions into a cohesive team effort. Co-ordination is concerned with ensuring that the output from each team member is directed towards the client's objectives. In order to achieve these, information must flow from one team member to another. Furthermore, conflicts between the various functions need to be resolved.

When considering links between people in a team, it is useful to compare the mechanistic, hierarchical view with the human relations view of teams in action. In the traditional view, each additional member of the team adds only a marginal load to the system. This can be seen in Figure 3. This shows that a closely managed team, with

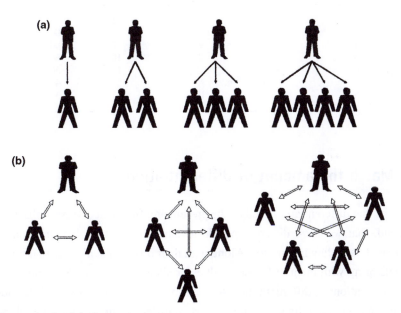

Fig. 3. Communication links in teamwork.

everyone reporting to the boss, means that each extra member adds a smaller proportion to the whole (a fourth person increases the number of links by a quarter, a fifth by a fifth). A more effective approach is to ensure that each and every team member communicates with the other, more in line with the human relations view. This reduces the demand on the project manager, but increases reliance on each of the team members. It also means that an additional team member adds a much larger burden to the communication network. This is why co-ordination and integration of large teams is so much more difficult than they would be for small teams. The changing demands on such a structure are shown in Figure 3. Although the team structure reduces the demands on the manager of the team, it increases the pressure on the information system and the need for careful design of integrative mechanisms.

Counting these communication links shows why each additional member increases the overall demands for integration so much. The relative increase in links brought about by each additional person is shown in Table 2.

Table 2. Relative increases in the number of communication links for different styles of organization

	Central control		Team work	
No. of people	No. of links	Increase (%)	No. of links	Increase (%)
2	1		1	
3	2	100	3	200
4	3	50	6	100
5	4	33	10	66
6	5	25	15	50
7	6	20	21	40
8	7	17	28	33

1.3.6 Match the amount of differentiation

The reasons that differentiation needs to be present are: first, that diverse skills are needed, and second, that we *want* different people to have different goal orientations. This is critical in understanding integration. Along with their different skills and goal orientations, individuals bring with them different loyalties or allegiances. Also, they may be working in different locations at different times. As mentioned above, Thompson (1967) identified technology, territory, and time as being important differentiating factors and allegiance as a reinforcer of differentiation. Allegiance may be to a firm or to a profession. The task of management is to ensure that there is the required level of differentiation in terms of skill and orientation, then to match this with a corresponding level of integration.

1.3.7 Co-ordinate

There is a variety of techniques for co-ordinating work and thereby achieving integration. The techniques presented here are based on the work of Galbraith (1973) who developed a useful scheme that was subsequently extended by Bennett (1992).

- *Procedures* – Procedures are standard ways of doing things. Establishing a procedure enables work to be done without reference to co-ordinators or a higher authority. Manuals and guides can be used to indicate the correct approach. Examples are the Co-ordinated Project Information Initiative, and Standard Methods of Measurement.
- *Hierarchy* – Managers are in charge of people and will therefore act to co-ordinate the teams under them. Co-ordination is achieved because information flows up the hierarchy so that the manager can take decisions and feed instruction back down the hierarchy. Clearly, this is critically dependent on the management information system. Different levels in the hierarchy have different levels of competence. If a manager is unable to deal with a particular issue then it is referred up to a higher level of authority. Integration and co-ordination is achieved when a manager straddles any discontinuities in the process. Such a manager collects information from one group and passes it to the next, while remaining in charge and monitoring progress.
- *Planning* – By planning and scheduling the work to be done, the involvement of managers in the day-to-day detail of the work can be reduced. If the work can be scheduled and made more predictable, then better use can be made of resources.

These first three are generally found in organizations and are associated with the desire to standardize and control internal operations in order to facilitate interaction between inputs, outputs and boundary transactions (regulation with the environment) (Dawson, 1996, p. 93). There are two further choices when the degree of complexity and uncertainty are high. First, the organization can attempt to reduce the amount of information required, and second, it can endeavour to increase the coverage and efficiency of information acquisition and transmission (Dawson, 1996, p. 94).

- *Creation of self-contained units* – Rather than a co-ordination device as such, the identification of sub-projects is a strategy to further reduce information processing demands. Since the difficulty of managing projects is directly connected to the size of the project, it may make sense to split a project into a number of sub-projects.
- *Slack resources* – Slack resources enable managers to meet their targets and reduce their dependence on higher levels in the hierarchy. It is difficult to organize work so that everyone is always fully occupied, and the cost of organizing it in this way may prove more than the cost of idle resources. Therefore, it makes sense to build some slack into the system. This includes buffer inventories, backlogs and time delays. One problem with this is the danger that this approach may be perceived as a general reduction in standards.
- *Environmental control* – Dawson (1996) suggests a range of options for organizations that wish to take a greater part in controlling the impact of the environment on their activities. Direct influence ranges from informal approaches, such as taking influential people to lunch or other forms of hospitality, to more formal approaches

such as inviting influential people onto the board of directors. Acquisition of suppliers and distributors is another means for attaining control over inputs and outputs. Bargaining and exchange with key players is another means, such as the way that property developers deal with town planning issues through the technique of 'planning gain'. A further option consists of using coalitions and cartels to achieve price stability. Contractual means may be used to secure ownership of ideas, or to assure delivery dates by providing mechanisms for penalizing suppliers who are late. Technological developments may have a direct impact on the relationships between certain competitors, suppliers and purchasers. For example, the relationship between Microsoft and Intel is very strong, even though they are independent firms. Marketing and promotional activities are an essential element of creating competitive advantage and uniqueness in the eyes of customers. If successful, suppliers too will wish to be associated with a firm and will therefore be more likely to give some kind of preferential treatment. The point about these activities is that they provide a firm with the opportunity to gain greater control of its own environment. While involving work and resources, the aim is to reduce the uncertainty surrounding the activities of the firm.

An alternative to reducing the information needed is to increase the information processing characteristics of the firm.

- *Task autonomy* – When management hierarchies become overloaded with information, one solution is to increase the autonomy of subordinates. One way to do this safely is to ensure that subordinates are members of a professional institution or a recognized craft that has a distinctive body of knowledge and, in the case of a profession, codes of conduct. Professionals and craftsmen bring with them a technical know-how as well as skill and discretion. By increasing the discretion and autonomy of those undertaking the work, less reliance is placed on the management information system. This works best when targets are used so that teams or individuals can judge the quality of their own output, rather than having everything checked by hierarchy.
- *Information systems* – The co-ordination devices so far discussed are sufficient for most simple situations. But when work becomes complicated, the demands for information increase rapidly. This can only be met by designing and implementing a management information system for the routine collection and reporting of information about progress, usually a prime aim of computerization in an organization.
- *Lateral relations* – An alternative method for increasing an organization's information processing capacity is to provide additional management roles with lateral connections. Through regular and direct contact people can co-ordinate progress and overcome difficulties as they arise. There are numerous ways in which such roles can interact.

1.3.8 Resolve controversy

The role of the 'co-ordinator' was analysed by Lawrence and Lorsch (1967a,b). Their findings had some very interesting points, particularly in the light of the current trends of emphasizing trust and goodwill (e.g. Latham, 1994). Lawrence and Lorsch identified the task of an *integrator*. They stated that such a person, in order to succeed, must be able to deal with conflict. They identified three methods for dealing with conflict; confrontation (defined as choosing, after discussion, a solution from those put forward for consideration), smoothing (defined as conflict avoidance) and forcing (defined as the naked use of power). They found that the most effective integration was achieved in organizations that used confrontation, supplemented as necessary by forcing behaviour to ensure that issues were properly confronted. Smoothing was the least effective.

While the vocabulary used by these researchers seems stark, it is clear that it is simply not enough to hope for agreement and harmony. On the contrary, if we employ an accountant, we expect such a professional to take an accountant's view, a view that ought to be different to that of an engineer. We want these people to bring their own agenda to any situation and to argue the case as they see it. Thus, if two people in the same organization simply agree, one of them is redundant (Tjosvold, 1985). Of course, although it is necessary for different views to be brought in and rationalized, the need for controversy does not extend to a need for dispute.

One of the most urgent tasks for any manager is the resolution of conflict and change. This 'takes up the largest single chunk of managerial time and energy, and is not always well done at the end of it all' (Handy, 1986). Conflict is a tremendous source of dynamism and creativity within any organization (Pascale, 1990).

Conflicting requirements will always need to be resolved in complex projects. Therefore, conflict cannot simply be dis-invented. Like cost, or time, the thing to be controlled is a resource to be expended as wisely and effectively as possible, not a phenomenon to avoid. The aim must be the resolution of conflicting requirements such that the project represents the best compromise from among the alternatives. This is the inherent nature of the design manager's role. Debate should be encouraged and controversy resolved. It has been suggested that the failure of architects to grasp this idea fully has precipitated approaches that circumvent the traditional authority of the architect (Gutman, 1988); perhaps the same is true of engineers. The need to resolve conflicts effectively and impartially is the same during the design stage as it is during the construction stage. The traditional approach has been to give the A/E the power to resolve all conflicts, thus ensuring that the design agenda (derived from the client's requirements) carries the most weight in all decisions. In this way, a comprehensive design philosophy underpins every single important decision on a project. This is a source of ambiguity in the role of an A/E under many construction contracts. The need to deal simultaneously with the role of employer's agent and independent certifier is exactly designed to deal with this situation. Unfortunately, too few people appreciate fully the reasoning behind this role and, worse, it is becoming increasingly apparent that not all decisions should be dealt with in this way.

As mentioned earlier (see Figure 2), the latest development in the progressive reduction of the role of the A/E is adjudication in which all challenges to the decisions of the A/E are referred to an adjudicator appointed for the purpose. While this move, to a certain extent, deals with the problem of role ambiguity, it also means that the agenda for such decision-making may not be primarily driven by design considerations. Under such circumstances, the design philosophy can easily become secondary. This is fine for strictly *contractual* issues, but may threaten the integrity of a design philosophy when such decisions impinge upon design issues. Thus, even with adjudication, the need to deal with design issues before awarding the construction contract becomes more, rather than less, important. Clearly, if the architect or engineer is to have a leading role to play, it is in establishing a comprehensive design philosophy for a project. This can be done by ensuring two things: first, that this philosophy underpins every design decision on the project, and second, that all of the decisions are taken before the contract is awarded. While the latter is a construction project management imperative, it is rare that a project is fully documented by the time of tendering. This is why the skills of a good conflict manager are as applicable to contract management as they are to design management. It is also the reason why it makes no sense simply to replace all A/Es as project leaders by some other arbitrary profession. Architects and engineers have collectively failed in their duties in discharging contracts, as many lawsuits have shown. Obviously, the professional title of the person exercising the main integrating role is of little consequence. What matters is the skill and experience that such a person brings to the project, and the framework of criteria established for effective management and control.

It is clear from this argument that the UK construction industry's approach of institutionalizing roles so that they are fixed for all projects, is far from adequate. In fact, it is probably damaging the interests of many clients. Such institutionalization may be appropriate in a time of little change, where most projects are similar. But that is not the situation that currently exists in the UK.

1.4 Approaches to collaboration

One very important theme in dealing with the interaction of designers with each other, with their clients and with those who build, is the way in which collaboration is brought about. The key issue is that teams are typically made up of people from different organizations. As stated earlier, the reason that there are so many different people involved is based on the complexity of the task and the consequent need for differentiation. But the reason why these participants come from different firms is connected with a branch of economic theory called transaction cost economics.

1.4.1 Cost of transactions

Transaction cost theory holds that the relative costs associated with different forms of economic organization are a major factor in the decision about whether to make or buy supplies. Coase's (1937) pioneering theory of the firm holds that if the costs of in-house production exceed those of contracting out, the firm will opt for outsourcing. The structure of a firm is regarded as the result of a competition between the price of internal resources and the market price. Construction and design work are almost invariably outsourced, except in the case of very large organizations. Transaction cost theory was developed and more comprehensively articulated by Williamson (1975), who, among other things, suggested that as well as the two polar archetypes of markets and hierarchies (buying and making, respectively), there was an intermediate structure called networks. It is this intermediate structure which is so well represented in the construction industry where the choices about which firms to invite to take part in a project are often dominated not so much by least price as prior knowledge and shared experiences. There has been a wide range of studies that have further developed the notion of inter-firm networks, although few writers have explained the design and construction processes in these terms.

1.4.2 Partnering and strategic alliances

One area where the notion of collaboration has been explored is Partnering. Several important pieces of research have sought to examine how the design and construction processes might be improved. Notably, Bennett and Jayes (1998) illustrated the processes of partnering with a series of case studies which showed how various partnerships in the construction sector have developed, with a view to developing guidance for best practice. This work followed from Latham's (1994) review of the UK construction industry in which calls were made for the investigation of partnering.

In other industry sectors, the development of strategic long-term relationships between firms has become known as strategic alliancing. An interesting phenomenon that has emerged in practice is that, even if there was no specific intention from the beginning, many strategic alliances result in one of the partners taking over the other (Bleeke and Ernst, 1995).

1.4.3 Supply chain management and inter-firm networks

The realization that longer-term relationships are important in bringing together successful project teams has also encouraged various people to think about their suppliers and their relationships with them. While the widespread movement of the 1970s and 1980s was customer-care, there has been a growing realization that successful project management requires supplier-care. Moreover, absence of integration between designers and the suppliers of materials and components can cause enormous problems with quality, functionality and liability.

The purpose of supply chain management is to build upon the notion of quality in the delivery of service, which has developed so strongly from the customer-care idea, by focusing on supplier-care. This is the underlying principle behind partnering, but with a more systematic and focused means for achieving the ends. The goals are consistency, high quality, economy, and satisfied customers. The means are to trace the supply and production of raw materials and components back through the supply chain to develop allegiances and information systems that enable effective development of the materials and components that will be integrated into the building. Figure 4 shows how the supply

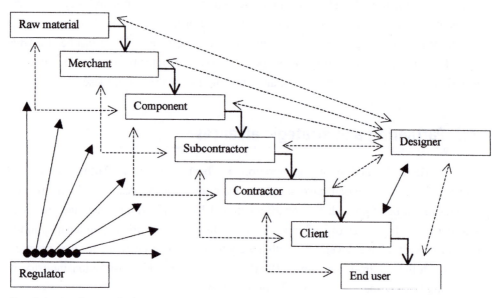

Fig. 4. Penetrating supply chains.

of materials and components can be seen as but a small step in a chain of transactions that are controlled by regulators and specified by designers. The scope for development of the principles of supply chain management is in developing dialogues at the links marked by broken lines.

By focusing on supply chain networks, firms contributing to the construction process will be asked to view their role as part of the overall process of delivering a building to an end user. This involves a fundamental change in focus and attitudes. Intermediaries in the supply chain will need to think about how their work achieves the aims of the building user, rather than simply the person who is paying for their services. It requires dialogue with one's suppliers' suppliers and customers' customers.

The challenge is to develop supply chains without compromising competition and free trade. A serious question is the problem of how we can structure the relationships and dialogues such that people along the supply chain will carry out their promises. This is particularly problematic in contractual terms because supply chain management may involve dialogue and promises between people who are not contractually linked (Benhaim, 1997).

1.5 Conclusions

Traditional approaches to the management of design are inadequate because they have evolved more slowly than the industry and society as a whole. The institutionalization of project roles has not helped consultants to develop more appropriate ways of managing the processes. Architectural and other construction-related education has been isolated from mainstream management thinking, from which there is much to learn about the difficulties of integrating and co-ordinating a fragmented process.

Organizational theory is the study of complex interactions between groups of people. Management is the study of how to get things done through others. Contingency theories of organization show that the best way to organize a complex task is to ensure that the skill diversity (differentiation of technology) is appropriate to the complexity of the task, and then to match the level of integration and co-ordination (management functions) to the level of differentiation. Organizational researchers have also shown that conflict and controversy are necessary elements in the web of interactions between people. Conflict has to be managed, not simply avoided.

The collaboration between individuals is part of the wider collaboration between firms in the construction sector. The construction industry is thus characterized as networks of transactions, a phenomenon that exacerbates discontinuities in the process, but an inevitable feature, given the nature of the tasks and the market.

The task of the design manager is to make sure that the organization of the design process is structured appropriately for the task at hand, and to ensure that there are sufficient integrative and co-ordinating mechanisms for the work to progress mean-ingfully. The connection between this and the design process itself is the subject of the subsequent chapters in this book.

2

The process of building design

Design is a creative and very personal activity. It is important, however, to understand how designers think when defining and realizing their objectives and their respective priorities. Only when the design is complete can the results of their intense intellectual activity be seen. This is at the heart of the problem of managing design. It is why managers of the design process must understand the methods by which a typical design is developed, and the characteristics of the designers, in order to achieve a level of understanding that allows them to be sympathetic to the process.

For the sake of clarity, only architectural design is examined in this section of the book. It is in this discipline that most of the research into the creative aspects of design has been carried out. Many engineers would disagree with this view and suggest that they have made an equal contribution to the creative design process (for example, Addis, 1996). While we recognize this, the characteristics of the creative process that are revealed and articulated here are applicable across the whole design team, not just to architects.

In essence, the architect takes the client's brief and uses design skills to develop a three-dimensional interpretation which other designers use as the basis for their own work. This is not a hard and fast rule, as on any project the formative or concept design stage is both interactive and iterative between the many design disciplines as well as between the architect and the client. The input from the structural and services engineers, for example, can often have a strong influence on the eventual design solution.

2.1 The characteristics of design

Design is primarily a personal task with the whole project's design becoming a combination of the motivation and expressions of many individuals. It is also viewed by the separate members of the project teams from many different directions and, in particular, with regard to how well it will accommodate their own needs and wishes. It takes time to explore, understand and consider the impact of these differing views. Any manager involved in the design process, must allow the necessary time for discussion and consideration, so ensuring that the designer's aims and expectations are met within the terms of the design brief. In this sense the design manager is in a supportive role, which allows the process to continue. The influence of a design manager cannot be ignored. This 'silent design' role (Dumas and Mintzberg, 1992) describes the decisions taken by non-designers who enter into the design process.

2.1.1 Design is a very personal statement of ideas

The final design is a culmination of a process that is often driven by personal motivation.

> 'Thus we can see that, although designers may be commissioned and briefed by clients and may chiefly concern themselves with the needs of others, the design process is also performed for the personal satisfaction of the designer' (Lawson, 1990).

What forms and influences the designer's character? What forms this inner motivation and how will it manifest itself in the designs that are produced? These are important issues to understand, not for the reasons used by many critics and writers to categorize architects by the styles they adopt, but rather to determine the nature and level of their commitment to their professed ideals.

Many designers, often driven by their inner convictions about the way the world should be, are determined to make a statement, whether political, social, monumental or aesthetic, through their work. This is developed during the process of architectural education, which in the UK at least is dominated by the project-based 'crit'. This system requires architects to develop their personal design philosophy and concepts and defend them strongly in open debate with their seniors and peers. There is a danger that this custom of vociferous defence may be perceived by non-architects as arrogance, but it is often so strongly developed as to be very difficult to modify and adapt.

2.1.2 Design is a form of art

Looking at a building should be a visual experience. Architecture is criticized and evaluated first and foremost on the basis of its contribution to the visual delight of the observer, either from the overall appearance of the external elevations or the way that space and light have been used in the interior. Great architecture is often likened to sculpture in the way that form is used to carry meaning, often abstracted from the initial concept and use of the building. However, architecture can seldom be the same as art in the purest sense of original creativity, because buildings are primarily designed in response to someone else's requirements. Perhaps in this way it is more akin to portrait painting, where the artist or architect uses representational skills to fulfil a commission.

'What distinguishes architecture from painting and sculpture is its spatial quality. In this, and only in this, no other artist can emulate the architect. The view that architecture is both a profession and an art is accepted by sociologists, but regarded by them as its chief eccentricity' (Pevsner, 1968).

This is not to detract from the skills of the designer, but perhaps to present a word of caution in that little about architecture is pure art in the conventional sense. There is obviously a great deal of creativity and originality required on the part of the architect in the interpretation of the brief. Indeed, an architect occasionally manages to achieve such originality that the whole of architecture is led in a new direction.

2.1.3 Design challenges the existing approaches

Much of design is concerned with using basic components and materials in new and different ways. It is by finding original combinations of products that designers can satisfy their own design philosophy. The range of technology available today, together with the rate at which it continues to develop, helps designers to realize their inner demands through the building design they create. Many new thoughts and ideas will occur during the lengthy process of building design. However, problems often arise from the fact that not all ideas can be analysed and developed simultaneously. This may raise the dilemma as to whether it is better to continue with the original idea, or to change to the new idea, which may well produce an improved design, but which may create disruption and uncertainty in the overall design process.

The need to satisfy the creative need is restrained within the confines of what is possible in the building process. There is an underlying tension in most architectural design. Set against a general aim of achieving good design, innovation and high quality is a constant pressure to reduce the price of construction. Therefore, a designer must set out the priorities in meeting these twin objectives otherwise ambiguity is inevitable.

2.1.4 Design is a realistic solution to a problem

In an attempt to understand the underlying methods used in the creation of a design, a great deal of research has been undertaken into the nature of the intellectual process used by designers. What is clear is that there is no single method or system used by all designers, nor does any one designer appear to use any single method (Lawson, 1994). A designer uses many methods simultaneously, directed towards solving the problem and arriving at an acceptable solution.

The creative leap is more a process of building a bridging concept between the problem and its solution (Cross, 1996). In an observation of the design process, Cross determined that there was an 'apposite proposal' from one member which grew in acceptance by the other participants in the group trying to resolve the design problem. Once the basic proposal had been accepted the whole group swung behind the idea and then put all its efforts into making it work. This is contrary to the more conventional view of designers waiting for the blinding flash of originality, although the apposite proposal has to be derived from somewhere. But even this was the result of an evolutionary process.

The strategy that appears to be used most consistently is one that focuses on identifying several possible solutions or hypotheses. These 'protomodels' (March, 1976) are evaluated and each evaluation is used to refine the proposed solution until an acceptable answer is reached. For this to be effective the problem must be clearly stated.

2.2 The nature of the building design problem

Clients cannot always state their requirements clearly or fully at the outset because of the many different interests that have to be satisfied. In many cases each problem is 'owned' by a group of people, each with varying requirements and ambitions for its solution. Each party will have a different role to play in the initial decision-making processes that will inevitably lead to overlaps and gaps in the statement of requirements given to the concept designer. To help develop a working brief that can be agreed by all parties, a designer will often pose solutions to the problem, largely to elicit where they fail and then, through a process of learning, offer a better description of the problem. External sources, such as Local Authority planners or environmental agencies, may well also have important inputs at this stage.

Designing is a process of human interaction and, consequently, the outcome contains the interpretations, perceptions and prejudices of the people involved. The acceptability of the outcome is also based on a trade-off among the individuals about what they are willing to accept as a satisfactory interpretation of their ideals. Inevitably, design is a trade-off between many conflicting needs until there is a solution that enables everyone to move forward to the next aspect of the problem (Akin, 1986), a process known as 'satisficing'. Each designer will have different perceptions of the problem and many ideas for its solution. Therefore, almost inevitably, features that one designer considers to be important may well be deemed unimportant by others. As Lawson (1990:89) put it, 'We should therefore not expect entirely objective formulation of design problems'. There may also be a cultural dimension from one country to another in the level of satisficing that is acceptable. This is a subjective set of criteria, but must be considered when a designer from one country is developing designs to be built in another country. If the designer is setting higher standards than the adopting country normally accepts then cost, technical and practical restraints may be introduced unintentionally into the project by the designer (Fowler, 1997).

A successful outcome to the design process is often determined by the choice of starting point in relation to the definition of the client's problem. Assessing the level at which to start is important and is a matter of fine judgement. It requires a clear definition of the boundaries surrounding the problem, because starting at too low a level may lead to a misunderstanding of the real issues. For example, a designer trying to design the layout of a floor space for office use must know how many people are to work in the office and the type of tasks that they are going to perform. This could then be extended to the way the office is organized and to considerations of the company structure and its philosophy on work, its workers, and many other matters. It is clear, however, that there would be no end to solving the problem if every eventuality had to be considered before taking any decision. At the same time it is equally important to allow the designer to stretch the boundaries of the problem to uncover all the factors that may influence the final design solution.

2.3 Strategies for solving problems in design

It seems essential to the design process that the designer should propose one or more possible solutions to the problem at an early stage, even if this is only to obtain a clearer understanding of the client's needs. Essentially, design is a cumulative strategy of developing a solution and critically appraising it to see whether or not it meets the criteria of the designer (Schön, 1983). More formally this can be expressed as: Generator – Conjecture – Analysis (Darke, 1984).

The designer may use a particular idea primarily as a route into the problem. This may be a constraint imposed by the designer in arriving at the solution and which can be so powerful that it may be retained through adversity and criticism until it either proves unsuitable or becomes dominant. This may apply to whole buildings or to systems within the building. It is this seeming reluctance to accept alternatives that is often misinterpreted as the intransigence of the designer. The conjecture and analysis stages follow and are largely based on reflection and review. It is a complex process aimed at increasing understanding through attempts to change or to reframe the problem.

An alternative view is that the design process comprises stages of interpretation, generation, comparison and choice, and the process is one of making a selection at each stage, as illustrated in Table 3. However, the proposition that design is a linear sequence can be questioned. It is probable that the designer thinks freely across and around the

Table 3. The four parts of the design process (from Hickling, 1982)

	The choice	The product
Interpretation	What is the shape of the problem?	A definition of the problem
Generation	What are the alternative solutions?	A range of alternative solutions
Comparison	What makes them different?	A set of comparisons and preferences
Choice	Where do we go from here?	A decision about policy and action

boundaries of a problem and a complex cyclic model is more realistic and representative of the process. This is illustrated in Figure 5.

There are two parts to this model: the iteration and evaluation within each part, and the iteration and evaluation between the parts. At any point it may be necessary to move to another part and evaluate it to understand the original problem, and so the process is one of continually cycling between and within the parts of the evaluation process. In practice this tends to fall into three stages: preliminary evaluation, probable solution, and final solution. Table 4 shows the many tools that the designer uses to propose solutions.

If this interactive and reflective approach to design is typical, then it makes a methodical and analytical approach to management difficult to adopt. This is because the

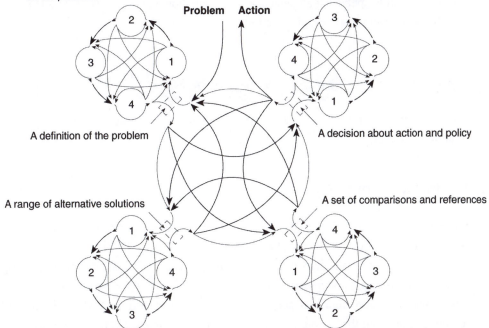

1 What is the problem? (Interpretation)
   1 What is this problem all about?
   2 What are the different ways of looking at it?
   3 Which ones describe the problem well?
   4 Can we choose one to help us get a grip on the problem?

4 Where do we go from here? (Choice)
   1 What are the pressures acting on the decision?
   2 What are the different ways of responding?
   3 Which ways will be effective?
   4 Can we choose what to do now, even if we have to leave some things until later?

A definition of the problem

A decision about action and policy

A range of alternative solutions

A set of comparisons and references

2 What are the alternative solutions? (Generation)
   1 What are the main areas of choice?
   2 What are the different solutions in these areas?
   3 Which of these solutions are feasible?
   4 Can we choose a range of alternative solutions for comparison?

3 What makes them different? (Comparison)
   1 What is the nature of the alternative solutions?
   2 What are the different ways of comparing the alternative solutions?
   3 Which of these provide accurate assessments of the relative merits of the alternatives?
   4 Can we choose a set of comparisons and references?

Fig. 5. The continuous whirling process of design thinking.

Table 4. Design techniques for the generation and comparison parts of the reflective process (from Broadbent, 1973 and Hickling, 1982)

Design solution methodology	Description
Pragmatic design	The use of available materials and methods without innovation
Iconic design	The copying of solutions
Canonic design	The use of rules and systems
Analogic design	Use of analogies from other fields
Lateral thinking	Ideas from another field applied to the problem
Brainstorming	Ideas, many random, focused on the problem
Models	Physical representations of ideas
Mathematical techniques	Detailed analytical techniques
The Rhubarb Principle	Moving forward to a solution before a rational linear process would allow
The Sutton Effect	Selective allocation of scarce resources to the area of greatest benefit

process is difficult to specify in advance and the evaluation against many criteria is likely to lead the designer in unexpected directions.

2.4 Constraints on the design process

At any point in the process the designer is working within a set of constraints, both internal and external.

2.4.1 Internal constraints

The internal constraints are imposed through wanting to work in a particular way, or with particular materials or technologies. These constraints may limit the range of solutions or may present the client with an appealing consistency. Therefore, certain designers develop a particular style of design or approach to the design problem and their reputation is based on this. Many of the constraints are self-imposed through self-criticism, which, if taken to the extreme, can mean that a solution that satisfies the designer can never be achieved.

2.4.2 External constraints

External constraints come from many sources, but essentially fall into four categories: the client's needs, the technology, the construction process, and statutory control. The requirements of the building itself will place constraints on the designer because of the need to satisfy the demands of the client. The nature and location of the site will also play an important part, as will any environmental conditions.

The medium of expression is through the technology of construction materials and products. While there is an enormous range of these with a virtually infinite degree of flexibility, there are practical limits to the performance of the materials available to the designer.

Further constraints are imposed by the methods of assembly of the materials and products. There are also practical limitations in terms of spans, weights and sizes of components that make certain options less desirable. Finally, designers of buildings do not enjoy complete freedom; various laws impose constraints on what they do, through planning, building control, and health and safety legislation.

These are the boundaries within which the designer must normally work, although some designers seem to be less constrained than others when their imagination allows them to make creative lateral leaps that identify new solutions.

2.5 The drawing process is an essential part of designing

The need to express designs through solutions and to communicate these clearly to others means that the process of drawing is inextricably integrated with the design process. In fact, design is impossible without some representational medium. The process of making the drawing requires that design ideas are developed and decisions made. Nowadays drawings may be produced on computers, and the way the images are generated greatly influences the mental processes of the designer. Whatever methods are used, they place limits on the range of designs likely to be envisaged.

2.6 Summary

While design has been dealt with in architectural terms, the same basic approach applies equally to the creative work of each of the other design disciplines.

Design activity is complex. To be able to manage the process effectively it is necessary to be sympathetic to the designer's ambitions and method of work. This can be difficult because of the following.

- The search for the perfect solution can be endless without constraints.
- There is, as yet, no perfectly correct process or solution.
- The process involves identifying the real problem as well as solving it.
- Design inevitably involves personal value judgements.

- There is no simple scientific approach to solving the design problem.
- Designers work in a complex and interactive way; this requires the additional focus of prioritization to the project delivery objectives.

The design manager must achieve the following.

- Allow designers time for reflection.
- Work with designers who have relevant experience and encourage and provide the support to enable them to find solutions to a problem.
- Establish a framework within which the tasks and objectives are kept in focus as the design moves through its stages of development.
- Provide access to the client for review and provision of more information.
- Help the designer understand the full implications of a new definition of the design problem and the possible need to re-enter the design cycle.

3

The process of engineering design

An engineer who doesn't care a damn what the design looks like as long as it works and is cheap, who doesn't care for elegance, neatness, order and simplicity for its own sake is not a good engineer. The distinctive features of engineering are mainly matters of content – the nature of the parts and the aims' (Ove Arup 1985).

The engineering stage of the design is where the detailed information from which the actual building is manufactured and constructed is generated. The output is the detailed, design information in the form of drawings and specifications that will be used for construction. It is primarily engineering applied within an overall architectural concept, involving the management of many designers.

3.1 The scale of the task

The greatest activity generally occurs during the engineering design stage when the design process is at its most complex, as well as being under the most time pressure. The number of people and organizations increases dramatically from those involved during the preceding stages and it is the high level of interaction between the groups that requires careful management. Nowadays the information from the trade contractors is vital for the development of the engineering design team's final drawings.

3.2 Design practice

The quality of many UK buildings is the result of careful attention to detail by the designers. The often-quoted phrase of Mies Van De Rohe that 'God is in the details' expresses the need to take the metaphors and style of the details and represent them throughout the building both externally and internally. This coherence of detail is very important to many designers and it follows that the design team is extensively involved in every aspect of the design process. In this respect, design practice in Britain is unique, whereas in the USA, Japan and Europe, much of the resolution of the final detail is left to technical staff and specialist contractors.

Once beyond the conceptual stage of the project, the scale of the design management task depends upon the designers, the choice of available technologies, size of the project and its environment. Any requirement that goes beyond the normal boundaries of the craftsman's skills will require specification and detailing. Modern buildings no longer depend on the normal application of craft skills and so the quantity of drawings and production information required is increasing.

The technologies of construction have fundamentally changed in recent decades for five reasons.

- A better understanding of the science of materials.
- Improved manufacturing processes.
- More competent designers.
- A willingness to challenge the conventional application of materials for aesthetic, cost or production reasons.
- Globalization of the knowledge and availability of materials.

In this book it is sufficient to consider two distinct applications of construction technology to gauge the effect on the production of design information: craft-constructed technology and engineered technology.

The difficulty of managing the design process will increase with the move away from the simple and well-known craft-based application of traditional materials towards the highly engineered and innovative use of new materials and technologies. In many cases, the site operations will not be affected in the same way as the fixing skills are relatively constant. However, greater guidance will be necessary to achieve the required application and finish.

3.2.1 Craft-constructed technology

Where the designer is content to allow the material to be applied by the average operative who is able to achieve a good standard of finish, there is little need for a large number of drawings or details. The same applies if the designer is content to allow the design intentions to be interpreted by skilled craftsmen.

3.2.2 Engineered technologies

Progress in building technology and types of contract determine the way that design information is produced. The design team will have to produce far more detailed drawings if the craftsman is not able to take part in decision-making. Where the requirements for fit and finish are higher than normal, additional specifications and details will be required. Also, where crafted materials join the engineered components, it may be necessary to specify and detail the junctions more fully.

Under most building contracts the contractor has no responsibility for design. Therefore, bearing in mind the penalties or liability the contractor may incur if work is carried out incorrectly, the designer will be expected to produce details that leave no doubt about the design requirements.

3.2.3 Component-based technologies

A fundamental change that has occurred in the last decade in the construction industry is the increased use of pre-fabricated manufactured components. There are three varieties in use.

- *Standard components* – Manufacturers offer a range of standard components in catalogues and it is simple to order and fix them in accordance with the manufacturer's instructions. Where they are to be integrated with other components it may be necessary for the manufacturer to issue specific manufacturing details for review or information. The recent introduction of safety regulations to cover hazardous substances has also made both the designer and contractor more aware of the precautions that have to be taken in the handling of substances that can be harmful to operatives.
- *Adapted components* – Many suppliers offer a basic range that can be adapted to the specific requirements of the designer. The manufacturer will produce a detail of the revised product for approval before the manufacturing process is started. There is often a second stage involving the production of shop- or manufacturing drawings that are then sent to the designers for review.
- *Specially developed components* – Many components are initiated by the designer and then specially developed in conjunction with a specialist supplier. There may be extensive research, development and testing before the details are accepted. The specialist will then undertake the preparation of shop or manufacturing drawings that will be sent to the designers for review.

3.3 Technology, quality standards and information

The complexity of a design is determined by the number of technologies used in combination, and by the degree to which the craft operative is left to interpret the requirements. Complexity increases with the variety of manufactured components used and accuracy of fit that tightens with the reduction in tolerance. These factors increase the volume of drawn information required and consequently there is a need for a higher level of design management (see Figure 6).

Manufacturing industry is also addressing this issue as more products are differentiated for increasingly specialized market niches. Even within the more tightly controlled manufacturing process the true costs of providing variety are not always understood (Martin and Ishii, 1997). The current goal of manufacturing is mass customization (Pine, 1993) or the adaptation of products to suit specific customer needs, while keeping costs at or near to those of mass production. The enablers of this approach are maximization of user-friendly IT systems, rapid change of production techniques and equipment, sophisticated suppliers and supply systems, and new management structures. Building is also a process with an increasing need to deliver a customized solution, at a lower cost and in a shorter time than before. It is only the scale of the product that is different; the issues are largely the same. However, techniques that measure the 'amount' of variety in building design have yet to be developed. Within the 'Design For Variety' initiative in the USA, tools are being developed which combine indices for commonality (or lack of) of components, differentiation of components, and the additional time to switch manufacturing techniques, into one quantitative tool (Martin and Ishii, 1997). The relevance to design is that the selection of the technology to be used in the building and its adaptation leads to the same issues of variety and also the same problems of identifying the true costs, both in designer's time for design and the consequences on the process.

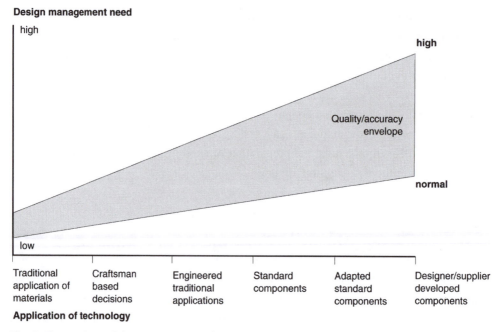

Fig. 6. The envelope of the increasing need for design management.

3.3.1 Design is a network of tasks

The complex design process comprises contributions from many specialists. Time passes between each transfer of information while it is assimilated into the recipient's own design generation process. The time taken to incorporate the information depends on the amount, its quality and its importance to the generation of the total design. In many instances design work in separate organizations can be done in parallel, but because the aim is to achieve a single integrated object, i.e., the building, there must be substantial cross-referencing between each specialist designer. The work content within each specialization is likely to be different, affecting the resources required and the time taken to complete each section of the work. Each exchange may result in positive progress or it may cause changes to be made to previous stages.

The design process, therefore, is an inter-related and uncertain exchange of design information between design specialists, which requires careful planning and co-ordination. When the design process for a building consists of a small number of sequential operations, then the management of the whole process is relatively simple. However, because each component is fixed to others, there have to be frequent and detailed interactions between the designers of each component to ensure that the fixing and support provisions are compatible. The sequences and interfaces between specialist designers thus form a network of design activity.

3.3.2 The environment for developing effective details

The link between the simple application of robust technology and cost effective construction is widely known. Simple systems, which are used without major modification, may not produce exciting design, but they do allow repetition, learning, and fail-safe execution. In the USA the specialist contractors have a wide scope to use the same approach from project to project. Consequently productivity can be high, as can wages and consequently the general social standing of the workforce (Flanagan *et al.*, 1979). The current urging to use standardized products is driven by the recognition that they lead to economies. Both the Japanese and American systems allow the use of standardized products across projects. This is not the experience in the UK where the fear is that standardization also leads to sterilization, and clients, in particular, do not want it (Latham, 1994; Gray, 1996).

The contractors in both USA and Japan have realized that construction requires a stable environment if it is to be able to deliver a good quality product. They therefore take great care to manage the relationship with their clients. As construction takes a considerable time the business requirements of the client may change and a revised briefing to the construction team is given. If this gets out of hand then the construction is in a volatile environment that is very difficult to control. A successful project is managed by taking control of the client's decision-making processes as early as possible to provide the certainty of decision-making. This is usually done by totally involving clients in the detailed decision-making process so that they can appreciate the necessary timing and consequences of the decisions.

In Japan, the learning is encapsulated within the contracting organizations as they, essentially, operate as design-and-build contractors once the outline design is complete. In the USA the contractors manage the specialist contractors to achieve cost effective design. Over most of Europe similar arrangements are used during the detail design by specialist agencies, such as the Bureau d'Etude in France and Belgium, to create a production engineering approach. In the UK the design team performs this function, but is focused on achieving design excellence not necessarily production efficiency.

There seems to be a polarization in the approaches to design and construction depending on the prevailing direction in which the members of an industry wish to place their priorities. The UK places a very high priority on design through close attention to details. Other countries place a high priority on cost effectiveness through a close attention to details. Both systems are using specialist contractors to achieve these goals, but the effect on the process is very different.

3.3.3 Technology to reduce cost vs. design differentiation

To understand some of the implications of a design-led approach, a sample of seven UK buildings were analysed (Gray, 1996). The examination of the primary component sets shows that there is a very high proportion of one-off component use. At least a half of all components are unique within the buildings' design. The average number of times that the remaining components are repeatedly used within a design is between two and six. Although some of the structural components, such as floor bays and beams, do have very high repeats which may give an initial impression that the building has been designed for economy, there is no general consistency. At this level of unique component use there is little cost advantage associated with off-site prefabrication so the only gain is rapid site production.

3.3.4 Attention to detail vs. risk taking

A low-cost approach demands that there is close attention to detail and that this is continually scanned to remove unnecessary items, to ensure repetition and simplicity. An attention to detail requires the right information to be produced by the organization most able to optimize the design for production. In UK-based projects, the total number of drawings averages one drawing for every $9\,m^2$ of building. There is wide variance from project to project, but generally they match the benchmark of one drawing for every $10\,m^2$ (Gray, 1995). The specialist contractors produce, on average, one drawing for every $17\,m^2$ or approximately half of the total number of drawings. This is consistent with a study described in the manual that is the precursor of this book (Gray *et al.*, 1994). In the USA, where the contribution from the specialists is sought to achieve the economic advantages of production and cost efficiency, the specialists produce four times as many drawings as the design team (Flanagan *et al.*, 1986). However, the total number of drawings is a tenth of those found in the UK for an equivalent building.

One conclusion is that the UK design team is involved in the detailed description of every aspect of the project and the architect takes the lead in their specification. There is discipline, but it is to achieve the required design solution. Specialist contractors make a significant contribution in both systems. In the UK it is to contribute manufacturing and technical knowledge to achieve perfection in the detailed execution of the design. In other countries it is to use the same knowledge, but within a broader framework that requires the optimization of the specialist's inputs to achieve cost effective construction. This is an overall simplification, but it does have significant impact on the complexity of the design management problem during the detailed engineering stage.

3.3.5 The essence of UK building design

It is worth considering the characteristic approach of UK design as it applies throughout modern construction (see Table 5). With the internationalization of design and the increase in wealth of many countries there is a trend towards buildings that are more complex. UK

Table 5. The production implications of changes in design and technology

Architectural development	Production impact
Constructivism	Functionalism, standardization, prefabrication and industrialization
Modernism	Fragmentation, specialization and volume production
International modernism	Component based to achieve quality and volume
High tech	Machined components requiring high grade skills
Neo-vernacular	Traditional material skin architecture within a new technological environment
Post-modernism	Return to classical influences and components used to replace the lack of skill base

buildings combine a complexity of uses and a complex design approach that may influence others. It is a perfectionist approach to the details of a building that is combined with an arts and craft philosophy developed through many influences. The Victorians, because of the demands of factory space requirements, had experimented and perfected the use of cast iron structural components, but the remainder of the building used traditional craft-based materials and skills. It was not until the end of the Second World War that the UK had the opportunity to embrace much of modern architectural thinking developed at the turn of the century in Germany and Russia. The constructivist school, which developed in the 1930s, according to Ginburg, should disclaim all pretensions to style; it was a question of functionalism: of rational planning, modern constructional techniques, the standardization and prefabrication of parts, and the industrialization of the building process (Risbero, 1982). Much of this view is current today. It led then to a concentration on the function of the building and that the building form should reflect the function. This is turn places a demand on the designer to work closely with the client in determining the needs in great detail. This also is current thinking and practice. There was a strong link between social aims and design – developing high quality, mass-produced housing to iron out social differences. There was a strong belief that through high quality designs, the spirit and aspirations of society could be raised. In short, designers took on the responsibility for social engineering.

In the 1930s the Bauhaus school of design refined this approach. Gropius developed building designs as machines for living. The Bauhaus disbanded and the members moved to other countries to avoid Nazism. However, post-war opportunities to rebuild drew on many of their design philosophies as well as their skills. The style of international modernism was developed. This has been refined over the years based on the approach of Mies van de Rohe who recognized that designers could explore space and use it to satisfy their client's needs and that technology could solve the resulting structural and closure requirements. According to Risbero (1982), 'In general Western architects were using form to make a technological statement'. However, the technology can become an end in itself so that the ideals of Mies van de Rohe, 'the whole (structure) from top to bottom to the last detail – with the same the idea,' when embodied in the design can dominate (Frampton, 1994). The good architects recognized that, in following this approach, 'exposed engineering details and use of materials need to be immaculately hand crafted not to look crude' (Risbero, 1982). This leads to the attention to detail, which manifests itself in a perfectionist approach to every aspect of the building, whether it can be seen or not. This is a European approach typified by a German worker quoted by Peters when asked why the weld quality on the back of the machine needs to be as good as that on the front:

> It is not a matter of the customer seeing it or not, the idea is to present the ultimate in quality (Peters, 1992:295).

Quality here is defined as attention to detail. It may seem odd to take this approach to the same extreme in a building that is on a much larger scale, but architects such as Sir Norman Foster typically specify tolerances of 1.5 mm over a 9 m length for cladding components (Williams, 1989:61). In the case of the Hong Kong Bank building there was a 50 mm tolerance in the structure over the total height of the building (183 m). Once again the emphasis is on the carrying of detail through the whole design, as:

> 'the extraordinary capacity to see a single taut and faultless line through a building, from the minutest level of the way in which the metal for the handrail might be rolled, to the final shape of the building on the site. It was this unquestionable ingredient which alone translated these practical and unassuming buildings into realms of architecture . . .'

as Williams (1989:60) described Foster's approach in his previous designs.

In her *Guide to Recent Architecture*, Samantha Hardingham said that

> 'Among the clutter there is an architecture that is peculiar to its location, has been designed for a specific purpose and place in time, and is representative of an ideology beyond that of making a fast buck. Above all a relationship has evolved between the architect and the client with the shared pursuit of making a better building, not a nicer building' (Hardington, 1995).

This focus on tailoring buildings to meet the client's needs, which embodies the views on design expressed above, has been embodied in RIBA's strategy for the future, which was published in 1995 after a 3-year review of the profession (Duffy *et al.*, 1995).

Much of the theory developed in the 1920s is still highly influential in current design practice. There have been many 'periods' of modern architecture which are additive in their influence on current practice. Their combined effect has been to attempt to retain the essential craftsmanship of building work, but at the same time incorporate new materials and ideas from the industrial age. UK architecture can be summarized as follows.

● Serving clients by understanding and meeting their needs in detail.
● Getting it right through complete attention to detail.
● Always achieving perfection of fit, as in the crafted approach.
● Using technology as necessary to achieve the utility of the building.
● Building small, complex, buildings for flexible use.

This has led to a complexity of approach and design solution that has been affordable only by keeping wage costs low (Hutton, 1996). In some sense this is a luxury that other countries have not been able to afford except on prestigious buildings, whereas in the UK it is applied across all buildings.

3.3.6 Level of detail for control

It is often difficult to select the appropriate level of detail to be used in planning and managing the complete design of a project. Two clear points in the process provide extremes in the level of detail available: when the designer is appointed, and as the detailed design drawings are produced.

Both of these extremes are unsuitable for control, as the first is too coarse and the second is too detailed (Farrell, 1968). An intermediate stage between the two extremes must be identified to assess the network of interchanges and to provide a level of detail upon which it is possible to evaluate realistically the amount of design work and its progress. An effective planning methodology, i.e., the information transfer schedule, which can generally be applied from the scheme design stage, is proposed in Chapter 10.

3.4 Summary

The engineering design process deals with the creation of the production information necessary for site operations. The iterative nature of the detailed design process arises because it is almost impossible to produce designs that take account of all the factors 'first time round' and, therefore, there is a natural progression of the information from preliminary, to probable, to final.

The quality of the information at any transfer dictates whether the subsequent task can move to the next stage or whether it must be passed back to begin a new round of decision-making. There is an inherent assumption that the individual designers are capable of managing their own task, and it is the interfaces both within their own organization and with others that need to be managed in detail.

The management task for the engineering design process is, therefore, to identify.

- the information and design inputs that are needed.
- the knowledge of and information about components and systems that are needed.
- the interfaces between components.
- knowledge and information transfer at all levels of detail.

Management must obviously make sure that all the information transfers occur at the right time, which can only be achieved if the required knowledge and its availability has been organized and contracted. This has to be done within an overall management strategy, embracing every contributor to the design as well as recognizing the overall project needs.

4

Stages, roles and responsibilities

The process of design and construction is divided into stages, common to all projects, based on key sign-off points needed from the client (see Figure 7). These are: the approval of the functional brief; the approval of the scheme design; the completion of the engineering design and the placing of contracts either with a single contractor or by separate packages; and the final acceptance and hand-over of the completed project. Engineering is further subdivided: detail design by the architect, engineering and other specialist consultants within the design team; and detail design by the specialist contractors, e.g. workshop and fabrication drawings. The division into these sections is much coarser than the conventional Plan of Work contained in the Architect's Handbook (RIBA, 1970), but it has been chosen because it more closely reflects the stages of management needed to achieve a rigorous and disciplined regime of signing-off. It also allows a better understanding of the pattern of contributions to the design and construction process typical on most projects. Signing-off is a discipline whereby all the designers present to the client at the end of each stage a complete set of information that is bound into a single document. If it is accepted after review and the decision made to proceed, then the client and the designers will sign all the drawings as the agreed set and ambiguity of intention is avoided. These stages, together with the subdivision of the engineering design stage, will be used throughout this book.

4.1 Streams of activity

Across all stages of a project there are three main streams of activity that interrelate through the client's decision-making process.

Functional relationships

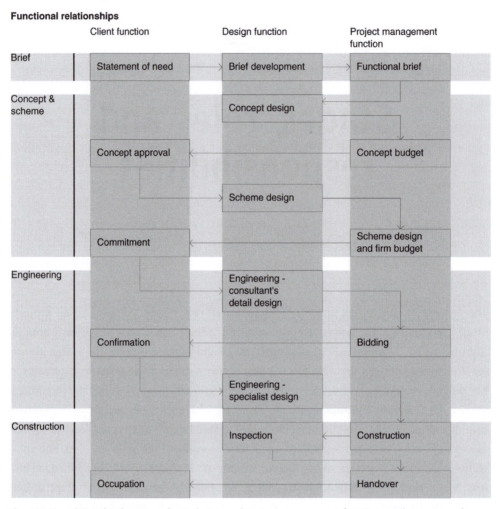

Fig. 7. Inter-relationship between client, design and project management functions in the process of design management.

- The client initiates the project and is required to make the decisions to proceed at each of the design and procurement stages.
- The concept designer assists in the development of the brief and produces the concept, which, if approved, is developed by the design team and other specialists into the scheme and eventually into the working and component fabrication drawings and specifications.
- The management framework enables designers and specialist contractors to work efficiently across the interfaces and ensures that suitable standards of time, cost and quality are established and adhered to during all the stages of the project.

4.2 Project and decision stages

4.2.1 Development of the brief

The client generally implements the procedure that leads to the appointment of the designers. Construction professionals may be involved beforehand, but the important starting point for the generation of design information is when briefing begins. The initial 'statement of need' must be prepared in terms that can be readily transformed into a design concept for development by the designer.

Experienced clients have a very clear understanding of their needs for specific building uses, and they produce very detailed briefing documents that set out criteria for the completed work environment, use of space, and quality standards. They are sometimes referred to as the 'scope' documents. Initial assessment and re-statement of the brief by the designer results in the preparation of a 'functional brief' document, which establishes the basic policy for the job. Most organizations bring this stage to a close with the creation and approval of a business case for the project that must be approved by the board of directors.

4.2.2 Conceptual and scheme design stage

From the statements of the client's need within the 'functional brief' the designer can develop the concept and outline design for the project. This is done with the project manager who simultaneously develops the budgets for time and cost. The design and budgeting processes produce a co-ordinated set of project information for approval by the client. Once approved, the next step for the designers is to work up the scheme design, where all the basic systems for the building are developed and checked for feasibility. Initial value engineering and buildability studies are carried out. The object is to fix the brief and the design solutions, including planning arrangements, appearance, construction method, comprehensive specifications, and detailed cost and time budgets.

The client must be satisfied that the scheme design meets the agreed requirements and should approve this, together with the budgets for time and cost. At some point between the briefing stage and the end of the concept stage, the budget ceases to be a consequence of the developing design and becomes a tool for control. On some projects, where the client imposes severe constraints, the changeover takes place at the brief stage. If project costs are to be severely controlled it is best to effect the changeover as early as possible.

4.2.3 Engineering

At the engineering stage, the design team develops the full production information. As the various systems are engineered and detailed they can be formed into packages to enable the project manager to start buying the manufacturing and construction work. The separation of the project into distinct packages is most commonly practised in construction management and management contracting procurement methods, with the content of each package dependent upon the form of procurement chosen. Where the package is large (e.g. the building frame or cladding), the level of detail prepared by the designers could be relatively small, particularly if there is to be a significant contribution by specialist trade contractors. Alternatively if the design is complex there will be extensive design development by the design team. Since these decisions have an impact upon the scale and timing of the project expenditure the client should be involved in confirming the purchasing proposals.

Once the specialist trade contractors have been selected and the contracts signed, they can start to produce their detailed designs. These must be incorporated in a co-ordinated way into the complete set of information for construction. This is probably the most complex and demanding stage of the whole design process.

The form in which information is provided for the construction stage must be appropriate for its use. The designers and contractors have a responsibility to agree their real needs together with the level of detail that is actually required. This concept is drawn from lean thinking (Womack and Jones, 1996), which means that the customer, in this case the production process, pulls the information from the producer, who is the designer. In practice this can lead to a significant reduction in information transfers because only the information that is really needed is produced.

4.2.4 The changing pattern of leadership

The level of involvement of different people within the project team changes from one stage to another, as does the leadership required to achieve the objectives of each stage. The differing levels of involvement (see Figure 8) are not absolute, but are relative, intended to show simply where the focus of influence lies between the teams. To ensure effective management at each stage it is important to establish and agree, from the beginning, the roles and relationships of all team members.

These broad stages in the design of a building remain the same irrespective of the form of building contract adopted. However, the stage when and method by which a construction-related influence is introduced into the design process will typically vary depending on the type of procurement system used. This may influence the roles and relationships between the participants, but the principles described here apply to any form of procurement.

Architect's plan of work - work stages	Services proposed	Design management plan of work - work stages	
A Inception	Brief and information gathering Site appraisal Advise on design work by specialist and other consultants	1	Brief: Statement of need Brief development Functional brief
B Feasibility C Outline proposals D Scheme design	Feasibility studies Outline proposals Scheme design Advise on effect of changes Planning application	2	Concept design, Feasibility studies and Scheme design
E Detail design F Production information G Bill of quantities	Detail design Cost checks on design Statutory approvals Production information Bills of quantities	3	Engineering: Detailed design by consultant design teams
H Tender action J Project planning	Other contracts Tender lists Tender action & appraisal Project planning	4 Procurement	Engineering: Detail design by specialists
K Operations on site L Completion	Contract administration Inspections Financial appraisal Completion Maintenance Record drawings	5 Construction 6 Commissioning & handover	

Fig. 8. Alignment of work stages to meet the Design Management plan of work.

4.2.5 The client's involvement

The client needs to remain involved to a varying degree throughout the process. As the design consultants are appointed and responsibility is delegated to them, so the client should normally expect less involvement with the project. This will vary with the type of project, the procurement approach, and the client's organization. The client's involvement is at its highest during the briefing, concept, feasibility and scheme design stages and will normally reduce during the detail design and construction stage (see figure 9). Detailed involvement with the building itself will increase again towards the end of the construction stage, as the client begins to consider the occupation of the building. At each

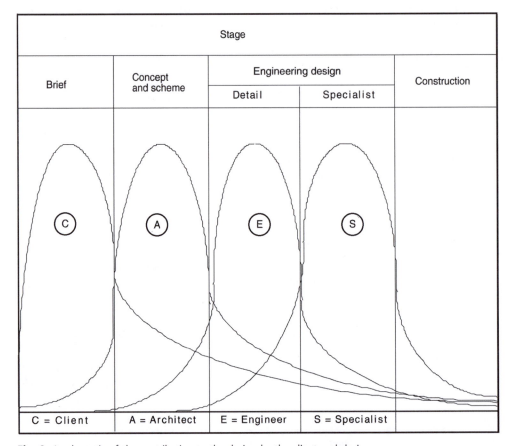

Fig. 9. A schematic of the contribution to the design by the client and designers.

stage the client must make positive decisions and commitments: the design and construction teams need this certainty if they are to deliver a satisfactory project.

4.2.6 The design team's involvement

The architect and the other consultants start to become involved at the briefing stage to help the client decide what is needed. The contribution of the architect-led design team is most significant during the concept stage through to the completion of the scheme design, and continues into the following stages through to construction. The engineering disciplines concerned are normally structural and services engineering.

The engineering involvement may be needed at the briefing and concept stages, it always covers the scheme and detail design stages, and will overlap with the procurement and construction. The engineers will usually be introduced during the concept design stage and as the project moves into the engineering stage their activity increases to a peak. They will also be involved through the next two phases of specialist design and construction.

4.2.7 The specialist's involvement

Although working information used to be mostly produced by the design team, an increasing amount is now produced by the specialist contractors who work with the design team to resolve the intricate details of the project. The specialists' involvement usually starts during the initial engineering stage and reaches a peak with the production of fabrication and shop drawings. However, unusual projects may well require specialist advice at earlier stages.

4.2.8 The construction stage

Design by all contributors will continue into the construction stage, because of the high design input by specialist contractors. The key objective is to ensure that the information flow does not interrupt the construction process. Therefore, once the manufacturing sequence has started within a product or work package, all the necessary information must be complete to allow the construction stage to proceed efficiently. There must also be a freeze on client decision-making and design development otherwise the penalty, in terms of extra time and cost, will be extremely high.

4.3 The 'wheel of dominance'

It is generally thought that the head of the design team is the natural leader throughout the life of the project. However, an examination of the pattern of work shows that different groups need to predominate during the design process at different times. It is clear that there are three distinct types of knowledge controlling the progress of the design, originating from:

- the client;
- the individual designers; and
- the overall management of the design and construction process.

The involvement of the individuals and groups who have this knowledge is rarely equal at each stage of the project. At any time one will be predominant and will naturally lead in decision-making and resolving problems. It is this creative tension which is the dynamic force driving a project. The primary concern is to achieve consistency of purpose throughout all phases of the project, which in the beginning is relatively simple as the client and design organizations only involve a few people and communication is straightforward. As the project develops many more people are involved and maintaining this consistency is more difficult, but it remains essential. The other task is to resolve points of conflict between the objectives of the client, designers and producers. This is shown (see Figure 10) by the addition of the 'wheel of dominance' to the schematic of the stages of work shown in Figure 9 (see page 48). For convenience three main groups are shown: clients, designers (covering all design disciplines) and consistent management (covering the over-arching framework required to manage the whole process effectively).

At the beginning, as described above, the client dominates and leads the briefing process. Once the briefing stage is concluded the scheme moves into the concept stage. The design team becomes the dominating influence, as it is the architect and the other professional designers who are providing the conceptual scheme, developing the

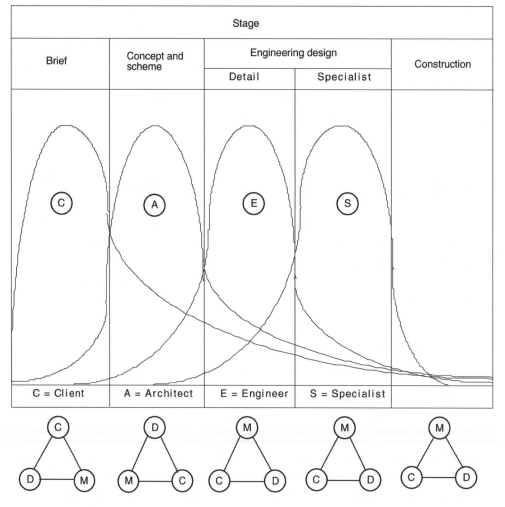

C = Client   D = Designers M = Consistent management

Fig. 10. The wheel of dominance.

aesthetics of the building, and making the main decisions at this point. Therefore, the tripartite wheel of dominance has turned.

The change that has had the most positive effect on the management of projects is the reconciliation of the roles of the designers and the manager at the engineering stage. Engineering and construction form the production stage of the project, where the production needs are paramount, hence the need for management to dominate and so the wheel of dominance turns again.

4.4 The network of management

There should be a consistent approach to the management of the whole design and construction process involving all the individual organizations. Ideally, the individual designers and technologists should manage themselves, but it is difficult for them to assess accurately their roles in the whole project. The task of co-ordinating the various project team members is normally delegated to project managers both within each organization and overall. The resulting network of project managers is crucial to achieve the successful completion of the design.

There must be an overall project design and construction management team whose responsibilities include the following.

- Creating the organizational framework for the design.
- Determining the programme and priorities of the design contributors.
- Co-ordination of the input of all designers.
- Evaluating the quality of the input.
- Process management that avoids, as far as possible, any medium or long-term problem in the supply of information.

Design management within each organization contributing to the design must comprise two levels:

- responsibility for the design and its production, with authority for decision-making on behalf of the organization; and
- responsibility for the interface with other organizations.

Thus a network of interaction through the project managers of each organization is formed (see Figure 11). These managers share the same project objectives and can take decisions in the commercial framework of the project to maintain continuity of production. It is a 'hands-on' role, rather than a co-ordinating and monitoring role. Nevertheless the operational parts of each organization may well communicate directly with each other to achieve rapid decision-making. This can be sanctioned as long as the relevant project manager is aware of and is involved in the decision-making.

Design Organization   **Specialist Contractor**

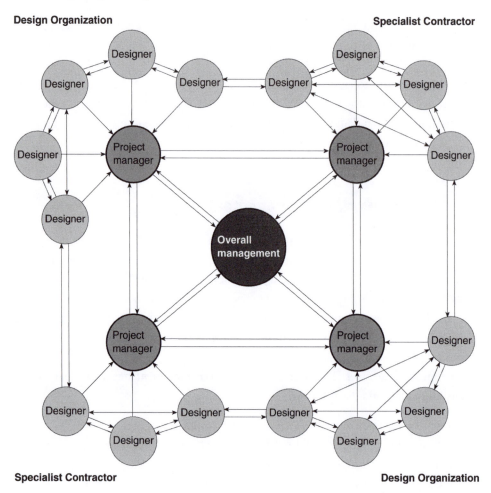

Specialist Contractor   **Design Organization**

Fig. 11. The network of project managers.

4.5 The effect of procurement options

There has been considerable experimentation with forms of procurement, with the objective of making the design and construction processes much more interactive. This has been due to the development of subcontracting, its increasing specialization and the need to provide a mechanism for designers to access the information. The development of various forms of contract and the implied relationships are shown in Figure 12. The most common form of contract is still the lump sum typified in the UK by the standard form of contract produced by the Joint Contracts Tribunal (1998) with quantities, but even this has gone through many transformations to enable specialist design information to be provided. In its basic form the design team does not have any direct link with the specialists and all communication is via the main contractor who, in many cases, will not

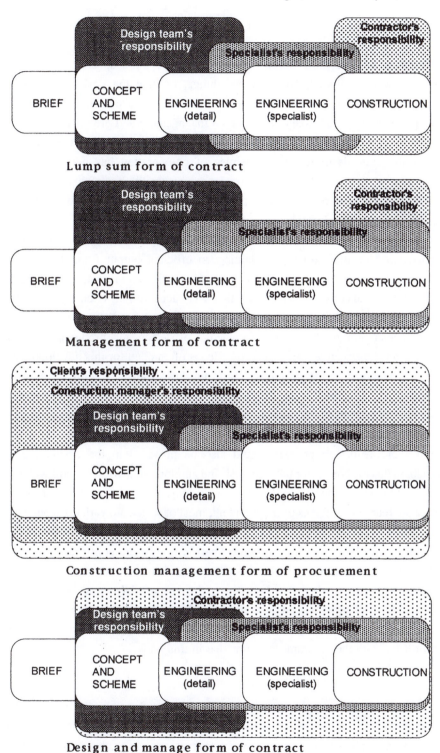

Fig. 12. The relationship between designer, specialist and contractor developed within common procurement arrangements.

accept design liability. This can result in a grey area of responsibility and liability as information is passed from one to the other.

As projects became more complex and knowledge more specialized, management contracting was developed, among other things, to give the designer access to the specialized information sources. This was done by the contractor becoming a member of the design team at an early stage to advise on management and production issues and, at the same time, provide an early purchasing system for specialist contractors. However, much of the enthusiasm for the principles of co-operation has been diluted as the management contractors have been made to accept many of the liabilities and responsibilities typical of lump-sum contracts and the division between design and construction has been reinforced.

A recent development has been the emergence of construction management by clients frustrated with the difficulties of obtaining an efficient output from the industry. By employing all contributors directly and providing the co-ordinating management, a cohesive design and construction organization is created, with the client assuming the total risk. One feature of this approach is the raising of the specialist contractor to a major position in recognition of the total knowledge of design and production within their specialization. American and European practice is often cited for this development, but it is a logical outcome of the developments in the UK. The major difference is that UK designers still retain considerable control over the final details and so the complexity of information interchange still exists, but within an integrated management-controlled regime.

Control of the design process would also be provided in the design-and-manage approach. In this the contractor takes the risk and employs all parties, but still with the same objective of managing the design process to achieve an efficient construction process.

There are many other forms of procurement, most of which are variations on these basic themes. All have arisen because of deficiencies in the traditional lump sum contractual system (Rougvie, 1981). The single biggest barrier to efficient construction has been the division between the designer and the contractor and the difficult problem of providing access to specialist knowledge without abrogating responsibility and breaching liability. Whichever form of procurement or contract is chosen, the parties should recognize the problem inherent in the implicit organizational structure and encourage sensible practice in accordance with the recommended practice in this book.

4.6 Liability

If the involvement in design changes so much from stage to stage, design liability cannot be easily attributed to an individual contributor. This is not a new problem. Every case of alleged negligence involves an enormous amount of auditing to attempt to apportion blame.

The roles that have been described represent the true picture of the management of modern medium and large-scale building projects. However, the extent of liability is not necessarily dependent upon who does what, but on who accepts liability. In other words, the documents used for appointing the various contributors are important in identifying the fundamental responsibilities for defects.

A client must decide whether strict liability is required so that the building achieves 'fitness for purpose', or that only 'skill and care' is to be exercised by the participants. The former is often more expensive, and is certainly more difficult to enforce as most, if not all, professional indemnity (PI) insurance policies will not cover 'fitness for purpose'. The client who wishes to pursue this course must make the appointments on a suitable basis. Alternatively, all parties in the project should undertake their duties with 'reasonable skill and care'. This is normally covered by their PI policies and therefore is enforceable. Each consultant and specialist must undertake to ensure that they perform their services adequately and sign a contract to carry out the work for which they have been appointed.

To clarify the position, there are two types of liability:

- professional liability, imposing an obligation on a consultant to act with 'skill and care', and;
- absolute liability, requiring 'fitness for purpose'. This may be attached to an agreement to provide a finished building.

The way in which the contract defines the work to be done will dictate the type of liability.

Tort actions are taken in common law and need no contract. The scope of this kind of recourse is restricted to physical injury and damage to other property only, and 'pure economic loss' is specifically excluded. As a result it is necessary to rely on contractual remedies, which is why collateral warranties have become so popular. They form a contract that defines the liability placed on the professional consultants and other designers such as specialist contractors. However, professional indemnity insurance policies usually exclude product liability, so consultants must be aware of what they sign. Clients must also understand the relationship between collateral warranties and PI insurance. A final word of warning is that each professional consultant must ensure that he or she is not the only one with a collateral warranty. The danger is that the consultant with whom there is effective recourse may end up with the whole of the liability.

The effect of the various liabilities and the difficulty of resolving problems through the courts has been that each designer takes extreme care to ensure that the information that is approved for construction is correct. Consequently more time is taken in checking and cross-referencing between the designer and, for example, the specialist contractor. Problems of co-ordination may appear where designers are extremely reluctant to sanction approval of each other's work and make commitments at the interface. On fast track projects, this raises considerable difficulties, as it may not be possible to speed up the design process unless sufficient time has been allowed to ensure that the design is correct.

4.7 Summary

The design process can be subdivided into many elements and stages. However, there are crucial points of decision where the design together with the cost and time budgets must be approved before the project can continue. It is the task of managers to identify these stages clearly and to achieve the following.

● Decide what is to be completed at the end of each stage.
● Ensure that the contributors agree to this.
● Ensure that the design team understands the change in relationships at the beginning of each stage in order to focus on the goals at the end of that stage.
● Ensure that the formal agreements with each of the design practices contributing to the design clearly specify the stages, tasks and changing managerial responsibilities.
● Ensure that designers, in attempting to limit design liability, do not confuse their co-ordination and management responsibility with liability for the content of their design and so limit their co-operation.

5

People and organizations

Although it is important to consider the structuring of relationships within a multi-organizational project, there are issues to do with collaboration and commitment that are just as important. Schrage (1990) offers three axioms of organizational design.

- Managerial work about the future happens linguistically, i.e., communication and collaboration are the essentials of modern work.
- The basic unit of work is commitment, i.e., a collaborating group must have a joint commitment to the goals being pursued.
- The basic structure of managerial and office work is the network.

This highlights the issue of assembling a team for a project. There are several steps that need careful attention; selection, management, communication, collaboration, and tactical approaches for turning these theories and ideas into effective project teams.

5.1 Selection processes

The most successful projects are often those in which the client has a long-term relationship with the designers, based on respect and trust. In the case of a client who builds frequently, these relationships are developed over many years and many projects. Empathy is eventually established between client and designer for the way that objectives are set, the way that they approach a project or building, and their attitude towards finding solutions to problems. It will take time and effort for a new client to develop this level of mutual understanding. While the relationship between client and designer is seen as creative, it is, nevertheless, very much a business relationship, with both parties having specific responsibilities to ensure that the project can be delivered. Therefore, the understanding between the client and each of the selected design consultants needs to be formalized.

Initially, informal meetings may be held between the client and design consultants to discuss the client's needs and objectives and how they may be interpreted, and to explore a preliminary strategy for producing a solution. Through these discussions a closer understanding and sympathy for each other's objectives will be formed. The designers must understand and explain any legal or financial constraints on the type and quality of building the client wishes to build, as well as its intended function. The client will want to know about the designer's approach to style and form and preferred types of materials, as well as ideas for the environment within the building. For this purpose the designer will often prepare drawings and models of the building to illustrate various options. Through this process the designers can express their creative talent and the necessary mutual trust will be built up. During this period, the designer's aim will be to identify clearly with the client the range of opportunities available for the development, and to establish the problems that have to be solved. Eventually the designers will identify themselves with the problems in such a way that they will own them and finding a solution will then become the focus for their efforts.

5.1.1 Selection of designers

After a number of potentially suitable design firms is chosen, the final selection must be made. There are many ways of achieving this, from competitive design proposals to competitive bidding. The client, together with any advisers, must decide on the method, but it should not undermine the mutual knowledge and commitment that has been developed up to this point. However, as this only deals with one aspect of the total selection process, it is also necessary to analyse the capabilities of the design firms.

The checklist for design selection in Table 6 is not exhaustive, but allows analysis of the four key capabilities of the design organization.

- Experience of designing similar projects.
- Experience and background of the people nominated for the project.
- Organizational resources and management systems.
- Financial resources of the organization and its ability to support the project.

Before conducting any formal interviews the client should establish satisfaction criteria and a scale of their importance, otherwise the resulting investigation may not be sufficiently rigorous.

5.1.2 Specialist contractor's contribution

Where the specialist is expected to contribute to the design or development of products, components and systems, the level of uncertainty must be recognized, and a means for apportioning the resulting risks must be obtained. The form of contract with the specialist should make a separate provision for the required design and detailing elements in order

Table 6. Project specific designer selection criteria (from Construction Industry Board, 1996)

Quality criterion	Key aspects	Suggested weighting range
Practice or company	Organization Financial status Professional Indemnity Insurance QA or equivalent system Commitment and enthusiasm Workload and resources Management systems Relevant experience Ability to innovate References	20–30%
Project organization	Organization of team Authority levels of team members Logistics related to site, client and other consultants Planning and programming expertise	15–25%
Key project personnel	Qualifications and experience relevant to project Understanding the project brief Flair, commitment and enthusiasm Compatibility with client and other team members Communication skills References	30–40%
Project execution	Programme, method and approach Management and control procedures Resources to be applied to the project Environmental, health and safety matters	20–30%

to avoid complex issues of liability and may have to be direct with the client. However, unless the mechanism for agreeing a price for the work allows for the development of the design, then both the design team and the specialist may become disillusioned.

The design team must be very clear about the level of design contribution expected from the specialist and must ensure that the specialist, the contractor and the client are also aware of this by precise drafting of all agreements and contracts. Equally the design team should be aware of how much the final costs may alter if extensive design development is done in conjunction with a specialist contractor. If this course is adopted, a rigorous regime of cost planning and control is also required.

Where the specialist is to be involved as an integral part of the design team, the selection process must be carried out early with a care equal to that used in the selection of the other designers. Selection procedures should give special attention to the most important elements of the specialist contractors' contribution to the project. The more the project depends on the individual technical and managerial contribution of the specialist contractor, the more their design, value engineering, and management capabilities should be scrutinized and weighed in the selection process.

The specialist's design agreement should be compatible with the design agreements of the consultants, the relevant activities scheduled, and they should be specified in detail. There are two principle ways for a specialist to contribute.

● The designer can produce sufficient information to determine the scope of the work and the contribution required from the specialist. The specialist can then give a price for the completion of the design and for carrying out the work.
● The specialist can make a specific design contribution and be paid for it. The subsequent work can either be carried out on the basis of a negotiated agreement with the specialist or can be placed by competitive tender with another.

Depending on the method chosen, the design and equipment selection process of the specialist contractor should be co-ordinated with similarly detailed activities scheduled for the consultants.

5.2 Management of the multi-organizational project

As shown in Chapter 4, the designers are one of three groups that form the total project organization. The three groups work together throughout the project, but within each group there are many organizations making a contribution. The effectiveness of the contribution is determined by the way that the contribution is organized in respect of the whole. The relationships can become very complex and even in the apparently simplest project, if the basic issues are ignored, then problems of misunderstanding and poor communication arise.

The view taken here is that the project must have a management structure that embraces each organization's contribution and places the people from that organization into a project-based structure that allows clear communication. This requires the managers of the project to understand each contribution, the individual contributor's organization, and the structure that will most easily accommodate good communication. A simple model (see Figures 13–16) can be used to describe the principles. The framework for this method of organization design is based on the analysis proposed by Jaques (1976). Within all organizations there is a structure of levels of work that are qualitatively different from each other.

The differentiation into levels is based upon two factors: the level of abstraction of the work that a person at a particular level is capable of considering and the time-span over which the person has control. The higher the level, the greater the degree of abstraction that can be comprehended and the greater the time-span of control. For people to be able to exercise the discretion required for the task, they need to have a manager working at the level immediately above their own. In other words, gaps in the structure of levels cause problems of comprehension and communication.

This model considers the relationships between the client organization, the design manager and the two design contributors; the design team and specialist contractors. The specialists are placed at the centre of the organization in order to reflect the growing importance of their contribution. The specialist's task is further divided into the two parts of their activity; design and assembly. The essential feature of this model is that there are no gaps in the hierarchy of levels and that communication flows vertically or horizontally. Therefore, a project organization that does not consider these issues will have gaps and communication will be flawed. In principle, communication flows either vertically between levels or horizontally across the same level. People in the same level generally deal with the same things at the same level of abstraction and understanding. Where there are gaps in the structure then communication is forced diagonally across levels between people who are at different levels of understanding of the issues. An extreme example would be the senior designer talking with an operative about strategic issues of the client's use of the building.

5.2.1 Managerial hierarchies in general

Jaques' (1976) work was concerned with the managerial hierarchies of production organizations. Here, this is applied to the work of design teams. As can be seen from Figure 13, the type of work that the people are required to do at different levels is quite distinct. Design teams generally comprise levels IV, II and I with a gap at level III. Also, individuals working at level IV may feel extremely frustrated when called upon to undertake tasks that require a lower level of complexity in decision-making, as for example, at level II. Level II work involves producing working drawings for translation into something to be built. The type of decision that needs to be made at level II is how to produce a clearly defined drawing that meets the needs of the site and adequately represents the designer's intention.

Level IV work involves combining an understanding of architectural principles, client requirements, financial constraints, site constraints and statutory constraints in an innovative way, to develop an overall solution or framework for construction to take place.

The work at these two levels is quite clearly distinct, the latter providing the context for the former. One person may find it difficult to do both types of task. Therefore, in a design team, people working at these levels should have their work clearly differentiated if not segregated.

5.2.2 Managerial hierarchies within a design practice

In order to understand the structure of a design team, and how to arrange relationships between participants, the managerial hierarchies in design offices need to be understood. Within an architectural practice the three levels of work are connected in two ways.

- Designers in the more senior positions need to manage the work of their subordinates, as within any organization, by allocating work and reviewing progress. This can be called Delegated Direct Output (DDO; Jaques, 1989) (see Figure 14). Tasks are broken into ever-smaller units and a product is delivered through the lowest level, in this case level II.
- Designers in the more senior levels are also responsible for their own output requiring the support of their subordinates for various aspects of the work. The work of the senior people is called Aided Direct Output (ADO) and that of their subordinates, Direct Output Support (DOS) (see Figure 14).

If these two sets of inter-level relationships become confused, the output from the design team appears confusing to those who have to use it. For example, to deliver the overall architectural solution, the work of a conceptual level IV architect requires input from the specialists in the various systems so that the proposed overall solution is viable. This can be from both level III system designers and from level II detail designers. The direct

Managerial work	Level	Designer's work
Senior executive Responsible for direction and representation of organization or very large project	V	*Senior partner* Responsible for representing practice and developing new technologies
Contracts manager Responsible for profitability of venture, including development of new and termination of old	IV	*Partner* Responsible for conceptual design and developing alternative solutions for client problems
Site manager Responsible for optimizing available resources to meet time and cost constraints	III	*Associate* Responsible for designing systems to make the concept work
Foreman Responsible for the allocation of work and the communication and overseeing of standards of output from Level I	II	*First entry professional* Responsible for detailed design of component parts
Craftsman Responsible for the direct physical output relating to the work of all levels above	I	*Technician* Prescribed output as defined by those in higher levels, e.g. drafting

Fig. 13. Levels of production organizations interpreted into architectural practice.

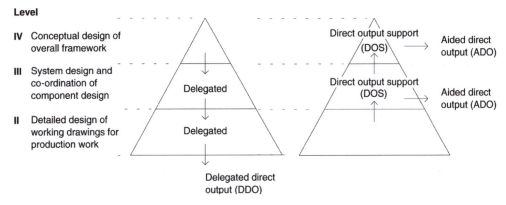

Fig. 14. Levels of work in architecture.

output support can be in the form of developed drawings, or models to check that the ideas will work in practice.

When a solution has been selected, the overall design has to be broken down into systems, and the systems into separate components for working drawings to be produced – the delegated direct output from level II. These are often not seen as distinct, and the two tasks, one for internal use and one for external delivery, get confused. The confused objectives can lead to the production of working drawings that are designed to meet the needs of senior partners rather than the constructors. It also means that the responsibility for the delivery of drawings from level II as a product is not managed.

Clearly the same people will be involved in these two sets of relationships, but they need to recognize the mode in which they are operating and, therefore, the nature of their responsibilities in each mode.

5.2.3 Communication between levels

In managerial hierarchies, individuals are sometimes placed too close to each other, producing a feeling of being continuously hemmed in. In design organizations people are frequently supervised from more than one distant level. This situation tends to create uncertainty, as the subordinate cannot grasp the full scope of the situation. Therefore, designers engaged in level III work are necessary to co-ordinate the conceptual design and the detailed drawings. This is described as system design work (see Figure 15). The work consists of ensuring that the conceptual design is broken down into the component systems, making certain that each of these systems functions, and that they are sensibly co-ordinated. On most projects the need for level III work is clearly necessary, but it frequently goes unrecognized. This failure is further compounded by the fact that the work of structural, mechanical and electrical engineers, and the input from the design teams of specialist contractors is primarily system design and needs to be co-ordinated at level III.

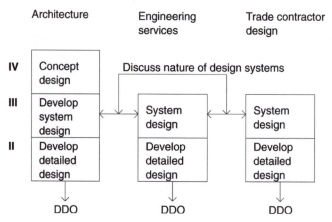

Fig. 15. Levels of work in a design organization.

5.2.4 Communication across boundaries and integration

Technologically-intensive work requires the input of specialist knowledge that must be communicated by people working at the same level and within a collaborative environment. Research into the type of communication roles that support collaboration, identified four types of boundaries within collaborating groups (Sonnerwald, 1996): organizational, task, discipline, and personal. The roles that are necessary to span the barriers were also identified. The key roles are two-fold. First, there is the intra-discipline role of linking within the organization to ensure good internal communication. Second, the role is one of inter-organization communication. These roles were identified by Lawrence and Lorsch (1967a) in their work on integration. The problem in construction projects is that there are subgroups within a multi-organizational project that form independently of organizational boundaries. Integrating roles are also required to operate between and within such groups (for example, see the discussion on technology clusters in Chapter 6).

5.2.5 Language

Language can be a barrier to achieving effective communication. One typical problem is the use of jargon or metaphors that are difficult to understand or relate to. Jargon is a useful way of encapsulating and conveying complex ideas quickly, but is only useful between participants who share the jargon. Architects, like any profession, have their own vocabulary, and can easily fall into the trap of using jargon with people who are not familiar with it. This is largely because they are dealing with issues that, while technical, have to be described in other ways in order to explain the underlying social and environmental philosophy.

5.2.6 Functional departments

The discussion on page 21 on the cost of transactions explains why many large client organizations have their own internal consultancies or functional specialists, and this adds further complexity to the managerial relationships in a design team.

Functional departments are generally found within client organizations that engage frequently with the construction process. These departments accumulate considerable expertise in many of the technologies to meet their (internal) client's business needs. In many cases this expertise contributes directly to the design process. A typical example is a civil engineer within a large utility. The relationship between the functional specialist and the project is fraught with ambiguity. The selection of specialist design input to join the other designers may be specified and decided by the functional specialist, who thus wields considerable power, on behalf of the client. The functional specialist may produce design information as part of the design process that seems to have the weight of direction from the client. While these are not actually client's instructions, such direction may possess more authority than it otherwise would. Therefore, the functional specialist's role must be dealt with carefully, otherwise both the client and project team may suffer. The stages of involvement of the functional specialist must be clearly identified and the differences in the developing role understood, as the project progresses.

Table 7. Changing responsibility of functional specialists

Project stage	Role	'Contract'
Briefing	Advisor as part of the client specialist information sources	Internal consultant
Design team selection	Task specification, selection criteria and assessor	Internal consultant
Detail design development	Design input, information generation	External consultant to the project team
Engineering design	Design input, information generation, evaluation of external consultants and specialist design	External consultant to the project team
Assembly	Quality standards	External consultant to the project team

As shown in Table 7, the role changes after commencement of detail design development from that of specialist advisor to one of developing production information, and the relationship needs to change to a formal one of external consultant with goals and fees clearly set out. This is only fair on the project team as the ambiguity of the role is removed and the relationship, and thus power of the functional specialist, is clarified.

5.3 Compatibility in the multi-organization

In practical terms, design work needs to be managed just as production work does, and the nature of managerial relationships for each project within the individual design practice must be clear and must reflect this need. The discussion so far shows why many firms and organizations are involved in various ways during design and construction. In order to deal with this inter-organizational complexity, it is necessary to achieve organizational compatibility between firms.

The creation of the appropriate design team organization involves exploring the connections between the different participants in the design process and understanding the way design work relates to the work of the specialist and construction teams.

5.3.1 Group size and problem solving

The term 'architect' is often used simply to describe the person who makes the decisions in a project. In practice, the majority of decisions are made in a group situation. Research has shown that group interaction contributes to an increase in the adequacy and accuracy of making decisions (Hollowman and Hendrick, 1971). Small groups of three wanted to agree on an answer and personality, persuasiveness and information available, all affected responses. Responses were not very accurate in very small groups, and with numbers greater than six there was no significant increase in the accuracy of the result. The conclusion was that a group with six participants would seem to be the optimum size for accuracy and efficacy in decision-making.

5.3.2 Managing inter-organizational boundaries

If boundaries between firms are to be managed successfully, and clear responsibility taken for the work to be done, instructions must not traverse level and firm boundaries simultaneously. That is, the connections between firms need to be at the same level of work, so that all individuals approach the problem at a similar level of awareness and decision-making ability.

When individuals at different levels are required to work in lateral relationships, there tends to be a great deal of misunderstanding between them. Those at lower levels do not get management and direction from their own higher levels, but inappropriate direction and demands from higher or lower levels in other organizations.

The schematic diagrams (Figures 13–16) referred to in the following sections, do not include everyone connected with a project, but are representative of the types of relationship that need to be established when creating a design organization.

5.3.3 Liaison with specialized design inputs

Although formal meetings between different design participants may need the level IV architect to provide the overall direction and framework, the primary communication between the design contributors needs to be at level III and perhaps at level II (see Figure 13, p. 62).

All firms need to be represented at level III because each has a specialized input of either overall systems, as in the case of mechanical and electrical engineer, or particular systems, such as the firm supplying the lift installations. Therefore, they have to be able to discuss the issue as colleagues all working in the same way. Equally important is that each would delegate work to the level II detailed designers who would be producing the production details.

Health and safety law often has a profound impact at several levels. For example, in the UK there is criminal liability for clients, designers and builders to ensure that competent people are involved in the process and to produce documentation that enables the finished building to be used safely. This requires a specialist in health and safety to be involved throughout the process.

5.3.4 Liaison with the construction process

Several inter-firm relationships may be explained by Figure 16. The one marked 'A' at level V, between the director of the client and the project director from the management function, involves sharing the understanding of what the project is going to achieve in terms of the client's business. It also establishes how the context needs to be created to allow it to happen. The overlap occurs because a project is being developed on behalf of a client. Therefore, a sense of direction is required from this organization. The project director can supply the particular expertise and can convert the client's expectations into an achievable direction. The critically important objective is to establish consistent management of the technical content of the project either from within the client's organization or by the client employing project management skills.

Level IV (marked 'B' on Figure 16) involves the people concerned with converting the overall direction into a set of specific objectives and ensuring achievement of the goals. It is here that those providing functional expertise, such as financial experts and concept designers, liase with those who will take general management responsibility for each phase of construction. The discussion between the people at this level will centre on what systems need to be developed and the resources that are to be deployed over the life of the project, recognizing that they cannot be aware of the full resource requirement, or how the conditions will change with time.

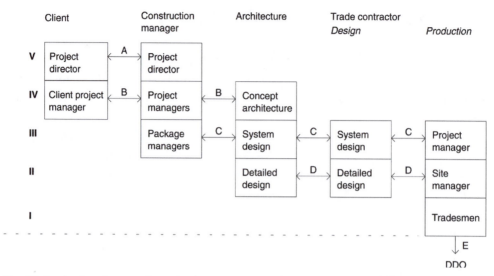

Fig. 16. Levels of work across the project team.

Each, through discussion with the others, will be seeking to provide colleagues in his or her own organization with a clear picture of the overall framework within which they will be required to operate during the course of the project.

The discussion between the designers at level III from the different firms (marked 'C' in Figure 16) is concerned directly with operational matters, such as how to make the best use of the means at their disposal, given the framework provided by the people at level IV. Given the nature of building design, this is where the interface between design and construction is most crucial. It is here that oversights in the design are made apparent and that the tensions between aesthetic and pragmatic concerns emerge. At this level the need for equal influence can be most clearly demonstrated. If some individuals were working at higher levels than others, then they would be able to exert undue influence in favour of their own concerns at the expense of other firms. At level II (marked 'D' in Figure 16) there is an overlap between the design staff and construction or production supervisors. The former deliver documentation that is to be used directly by the latter, who have to ensure that it is carried out to the required standard. At level I the design is realized and any confusion at an earlier stage will severely compromise their ability to work well.

5.4 Agreements

The formal agreements between the client and the designers (whether consultants or specialist contractors) must be carefully drafted with the client's objectives consistently set out so that the duties of all the design participants are compatible. Each of the professional institutions have developed their own recommended forms of agreement that have been drawn up to obtain the best from the client/consultant relationship. However, they are not co-ordinated and problems will occur if the implications of this are not understood. Care must be taken to ensure that the contracted service is compatible with the task that is to be performed by the consultant, and that limitations are not imposed on the scope of the service that might interrupt the necessary flow of information and decision-making. For example, problems can occur when one consultant is working with an agreement based on full duties and another is working on partial duties or a performance specification.

Much of this potential conflict can be resolved during the design planning stage when the design process for each of the building's components and systems are examined. At this point, the responsibility must be established for producing the information that is required to allow each stage of the production process to continue. It must also be decided which parts of the detailed design work will become the responsibility of specialist contractors, as this will affect the timing of the decision made on which form of construction contract will be used. Moxley (1984) provides excellent guidance on how to detail the designer's task for the purposes of negotiating a commission.

As they become more sophisticated, clients may seek to limit their risks, particularly those caused by failure of the designers to manage their tasks properly, especially where the failure results in claims for delay. The aim is to focus the design team on to their responsibilities. On the other hand, clients may seek to manage risk by retaining it instead of transferring it, as this may be more economical, depending upon the scale and frequency of the client's construction involvement (Murdoch and Hughes, 2000).

The two areas of concern are:

● any activity of the consultants that may directly affect the start on site, or the completion date, by failing to supply the information reasonably required by the other members of the design team; and
● the recovery of the costs associated with extensions of time granted to the contractor to cover the failure to supply information at the required time or because of variations required by the designer.

Any agreement between the participants ought to embrace the concept of changing leadership during the process, although none currently does so. This is a problem in all construction projects, but it is largely ignored. An ordered approach is necessary, which accounts for design leadership, liability and management. The agreement with each designer should reflect the contribution they are expected to make and the management framework within which they are to work. It should be clear at which stage in the project their contribution is required and who, within the stage, has overall authority.

The management framework is often implied through convention and practice, but as already shown, the conventional approach does not equate with what is needed. In the realization of the complete building, production is as important as the concept and the scheme design. The design agreement should be framed to allow an approach that combines design with the practical limitations of production against the constraints of the original objectives interpreted in terms of the criteria for success in the particular project. During the engineering and construction stages, the agreement must allow for the change to a production management-focused organization.

Well-organized designers will not leave themselves exposed to penalty through poor management of the design process, but others who are less well organized, may suffer.

5.5 Summary

Assembling the team needs to be done carefully. The team should be structured carefully because the site requires fully co-ordinated production information. This means a vertically integrated system design with controlled and co-ordinated interfaces. It is assumed that the detail of each system will be developed in conjunction with the specialist suppliers and trade contractors. Their input will be co-ordinated by the lead designer from the professional design team for the system.

Co-ordination will occur within the professional team, and between the team and the specialists, to achieve the technical requirements of the design. The organization and timing of the co-ordination should be agreed by both the design manager and project manager. However, they must ensure that communication occurs across levels and not between levels, to avoid confusion and frustration.

The nature and scope of the contributions of participants should be agreed and priced in writing before they start work. Participants internal to the client's organization will be obligated by their contract of employment. Those outside the client's organization will be obligated by the contract that is agreed with them for the purposes of the project. The disposition and balance of risk and responsibility need to be carefully considered in the context of the particular requirements of each client and each project.

Part Two

Design management practice

Tactics for developing a project culture

In developing an effective approach to the management of building design, the strategies outlined in the previous chapter form the backdrop for a variety of tactical approaches. These encompass communication, collaboration, technology clusters, location, and start-up meetings.

6.1 Effective communication

The issue of communication has already been addressed as an issue of organizational design where the structure of relationships allows good communication between the levels. However, this is only one aspect of the communication issue. The most common problem is the need to understand the network of communication and how it must be facilitated at each level. Most organization charts are two-dimensional in that they concentrate on the hierarchical issues and assume that the horizontal network issues are understood. Alongside these thoughts are those of teamwork and the 'design team'. We have tried to avoid the use of the phrase 'design team' because we believe that conventional team thinking is not appropriate for understanding how best to manage the design contributions. Design team is a loose collective team to describe the fact that the design process requires considerable input from a whole range of contributors. Most of the contributions come from a loose network of different organizations selected because of their knowledge, skills and manufacturing capability. Teamwork is assumed to be required and until a 'team' is formed the group will not function adequately. The view taken here is that the group needs to be integrated to ensure that it exchanges its skills and knowledge, but that within a temporary multi-organization of loose contributions, teamwork in the conventional sense will not occur. This is a subtle difference in terms, but has significant management implications because the assumptions associated with teamwork are not valid in multi-organizations.

However, given the above, the core design group of architect and key associated consultants, with their long-term involvement in the project, will build the characteristics of a team. These are long-term involvement, synergy of views, and an almost subconscious understanding of the reaction of other members of the team to proposals and ideas. This streamlining of work through the removal of unnecessary information exchange is the goal of true team working. In practice this takes upwards of 9 months before full understanding and trust occurs.

In practice, the project will generally generate a core group of people who will form the core 'team'. Others must be added to this group and will become involved for specific contributions to the project. The core team therefore has tremendous power through its long-term knowledge of the project and could dominate the contributions of others. Good management recognizes this and rebuilds the groups at key stages in the project to bring the necessary skills and knowledge of the project into play at the right time. This is considered more fully later in the chapter.

6.2 Collaboration

Given the time constraints and the practicalities of developing teams, a realistic approach to the needs of modern design is to aim for a degree of collaboration that enhances the creative and decision-making processes. Specialization is a response to complexity (Lawrence and Lorsch, 1967b; Schrage, 1990). Specialists cannot work in isolation, they need to be brought together to achieve a product that is the result of a combination of their skills. The task of management is to provide a collaborative environment to allow for the transfer of knowledge. Collaborators are constantly reacting and responding to each other and frequency of contact becomes almost as important as the nature of the contact. In practice, collaboration is a far richer process than teamwork. But the issue is not communication or teamwork; it is the creation of value. This is at the heart of the modern design process.

A chain is only as strong as its weakest link. In a technologically dominated environment the level of understanding of the detail of each technology must be as high as possible. This places pressure on providing the best people from the organization in to each group or increasing the learning of the organization to provide the necessary inputs.

Collaboration requires people to work together freely to the maximum of their potential. This can only happen where there is mutual trust and respect for each other's capabilities. Management must provide this type of working environment for each collaborating group.

Design problems are resolved through the evaluation of information inputs from a variety of sources. Therefore, successful collaboration must allow the continual exchange of information and knowledge without any barriers being put in the way. There should be clear lines of authority but no restrictive boundaries so that the communication can flow freely between organizations. This places pressure on the relationship between the formal and informal environments. The role of the individual organization's project managers as the custodians of formal communication cannot be underestimated.

Collaboration implies that information is transferred freely for use as appropriate. The lead designers in any specialization will be the final arbiters, although the collaborating group may have had considerable input to the assessment of the situation.

Construction projects require intermittent contributions, some of which last for a long time. Frustration is inevitable if people are expected to work as part of a team without adequate recognition of their contribution at the end of an intensive period. Collaboration must be understood for what it is and unreasonable expectations dealt with at the outset.

6.3 Technology clusters

Another view of managing complex and technologically-driven environments is to bring together all contributors to the development of the components at the systems level and ensure that the value chain remains unbroken from start to finish (Gray, 1996). The object is to merge together the work group with the support function into an integrated task group called a technology cluster.

'A cluster develops its own expertise, expresses a strong customer orientation, pushes decision making towards the point of action, shares information broadly and accepts accountability for results' (Mills, 1991).

The design and construction processes involve the combination of many components. Each component must be developed in a variety of ways depending on the nature of the component, from the simple manufacture of standard components, via tailored standard components, to specially developed components. The site teams can do little to effect the efficiency of this system as it is entirely controlled by the design and information production system.

6.3.1 Nature of technology clusters

The technology cluster, by focusing on the specifics of the technology within the manufacturing process, enables constant evaluation of every aspect of the process to ensure that the overall aims are not compromised by the desire to optimize a component.

The aims of technology clusters are as follows.

● Group all contributors together, preferably in one location.
● Elicit solutions to technical, quality and efficiency criteria to support innovative design solutions.

- Create a fully integrated systems-level solution.
- Focus upon completion of the system as an integrated unit.
- Preservation of the value chain throughout the supply chain.

6.3.2 Example of a technology cluster

This is illustrated in a typical technologically-complex area of a building, such as the substructure, by the pattern of involvement shown in Figure 17.

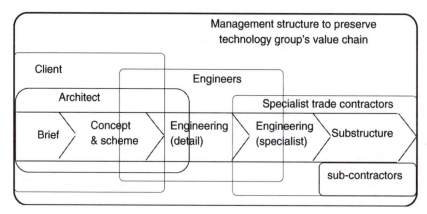

Fig. 17. A technology cluster.

The central process of developing the design for the substructure starts with the brief and involves the client and the designers. The concept and scheme design is developed together with the engineers to create the shapes, sizes and profiles. The engineers develop the engineering design in conjunction with the architect. They specify sub-components based upon this analysis. The specialist contractors undertake the specialist design in conjunction with the engineers to produce the detailing, e.g. reinforcement bar layouts. There may be a need for subcontractors to be involved and these are co-ordinated by the specialist contractors. To preserve the value chain, this whole complex group must operate within a focused, vertically-integrated management framework that supports the development of the technology. In practice, this chain is compressed with everyone making simultaneous contributions. The efficiency gains are made in two ways. First, by bringing the highest level of skill and capability to bear on the solution of the problem within a framework that challenges them to deliver against clear, project targets. Second, by doing this in an organizational framework that reduces communication lines to instantaneous face-to-face interaction.

6.3.3 Team building and location

A technology cluster is a multi-functional team. All the characteristics of good team building must be available and used. All components of the team must be present from the start. This requires an honest and open evaluation of the required knowledge and potential sources to be made at the start. A radical rethinking of the purchasing of the contributions is required, particularly the specialist contractors. They must be involved as early as possible if their capabilities are to be maximized, and ways must be found to achieve this. Teams need to be adjacent to production; technology clusters require close physical presence to work. In spite of the potentials of integrated IT there are, as yet, few examples of it working at this level so there is little choice but to put teams together in a location best suited to making project-focused decisions.

6.3.4 Skills

The people who are involved in a technology cluster need to be multi-skilled, with mastery of an area of special competence. The UK does not have enough organizations where professional knowledge is so exalted. A technology cluster requires that only people with exceptional levels of skill are involved. They must be given a freedom to work, so their work must not be over-planned. However, management expertise, both for the management of the group and the management of its contribution to the project, is always an integral part of the group. All members of the group must be able to transmit their knowledge to other people and organizations in the group. This requires good communication skills and an ability to contribute freely to the group.

6.3.5 Integration of technology clusters

A construction project requires several technology clusters, each working on separate systems within the project. Figure 18 shows that, while each technology cluster works independently, there is a need to integrate them. The purpose of this integration is two-fold. The first is to achieve the design and technological integrity of the whole project's design requirements. This is achieved by linking the architect and engineering roles between each technology cluster. The second is to maintain the subvalue-chains within a whole value-chain. This is provided by a management structure that embraces all aspects of the delivery of the project within one entity.

How is this different from a conventional approach to managing a project? It assumes a product focus to the sub-management structure. The product focus is to support and achieve innovative design solutions to satisfy the client's needs. Each cluster brings together all of the design and production skills in an intensely focused way. Focused communication by

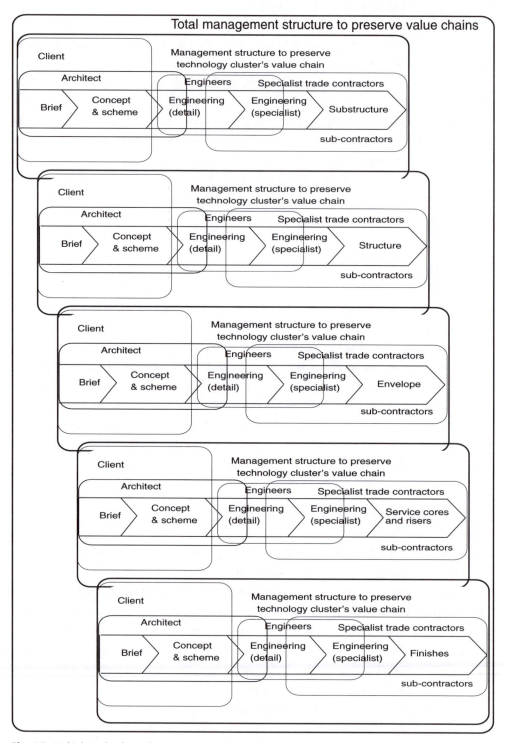

Fig. 18. Multiple technology clusters in a complete project.

physically adjacent people and groups of people speeds up the whole process. Furthermore, the management structure is very flat thus aiding decision-making.

6.4 Developing a unified team

A group that is meant to collaborate must have a shared and understood goal that unifies their activity. This requires commitment and regular, frequent contact.

6.4.1 Commitment

It is clear from Chapter 5 why the contributors to the design effort will be drawn from a variety of sources in many organizations. But, of course, each organization will have its own way of working and its own corporate objectives. These may be sympathetic to the project, but can often override it. Successful projects overcome these constraints and combine the contributor's identity with the project and its objectives. This can only be achieved through gaining the personal commitment of individuals.

The first team-building task is with the consultant design group. There is an assumption that because it is the originator of the design, there is an automatic commitment to the project by this group. In the concept stage this is easily achieved, but as the project progresses the number of people involved will grow and their association with the project will be different. Their acceptance of the same motivation as the originators of the design may take a considerable time to achieve, if it occurs at all.

Even more difficult is obtaining commitment to project objectives of the specialist contractors' designers. They are more remote and often involved only for a very short time. As long as the other participants understand the terms under which the specialists participate, then there should be few difficulties. However, if the contribution is large and a lot of people are involved, considerable effort is needed to avoid any tail-off in commitment.

6.4.2 Co-location and project design offices

A hurdle in achieving integration is that most design professionals work in their own geographically separate offices. Physical communication barriers have to be overcome by a variety of means. Many of these do not allow immediacy of the personal interaction of working adjacent to each other. It can therefore be very difficult to make the project priorities transcend the local priorities of each participant. To manage effectively such a diverse and dispersed organization, it is essential to focus the efforts of each individual on the needs of the project.

Co-location is a project approach to providing a site-based working environment where everyone is in the same building or space. The advantage is that integrated working is easier to achieve; communication is immediate and a genuine team spirit, that embraces the whole project, can develop. When this is enhanced by the presence of the designers from the key specialists and subcontractors, and a representative from the client's office, considerable savings in time and effort can be achieved.

If this method is chosen the project managers must understand that the designers need the support of their offices' information infrastructure, access to their peers, mentors and managers. There is a danger that they will become isolated from their base and from the conventional aspects of career development. If the site is located a long way from their home base, this will raise issues of travelling costs, absences from family and friends and the general social dislocation that this causes. However, when the costs of information transfer and communication under the conventional working system are considered, and the gains from rapid decision-making are set against the costs, the net advantage can be very large.

To establish a more formal approach to co-location, the client may choose, as part of the organization strategy, to provide a central location, a project design office, in which all contributors to the design process can work. On the occasions that it has been used, this approach has been successful. But it is not common.

A modification of this approach is to take the primary design team to a separate location away from the practice office for the crucial creative stages.

'Arup Associates' practice was built on a principle of having a place to go away from the office, where a core team could work intensively and uninterruptedly at appropriate times. I have also found it the most constructive and creative way to go about design. It is not only that you have the concentration and uninterrupted time, but also it is the fact that you can be involved intensively. It is to do with that intensity which can burn away superficially attractive ideas and so dig down to the essence of the scheme' (Dowson, 1990).

6.5 Start-up meetings

Even without a project design office, much can be achieved. It is important to identify, at each key stage in the process, the key designers and the personnel in their teams and to introduce them in a positive way to the objectives of the project. This is most effectively done at formal start-up meetings.

A start-up meeting is a meeting initiated by the client or the project manager (see Checklist 1) and differs in purpose from all other forms of project meeting. The principles are as follows.

● To introduce to one another everyone who is involved at the particular stage of the project, and to ensure that they understand their own objectives and those of the project.

Checklist 1. Action by project initiator

Appoint client representative

Select members of the working party

Circulate list of all members in the working party containing:

Name:

Position:

Address:

Tel. No.:

Position/Function:

Establish reporting and approvals procedure

Establish working party terms of reference

- Attendance of all the relevant people is essential. The project manager must decide who attends the start-up meeting, but there should be no barriers to attendance. If necessary, specialist contractors may need to attend in their own best interests as well as those of the project, and must be appointed in time to allow this to happen.
- The agenda involves a high degree of participation by those attending and is designed to produce a consensus interpretation of the individual contribution to the objectives and a commitment to them.
- The meeting lasts for one day, with no interruptions.
- Each party prepares a 10-minute presentation of what he or she sees as the objectives of the project and his or her contribution to it. This is reviewed and any differences are resolved so that there is common understanding.
- Once a common understanding of the objectives is achieved, a plan for the subsequent stages of work is developed and agreed. Each contributor must be prepared to make a commitment on the detail and timing of his or her contribution and explain what he or she expects from other designers. The planning is done in open session with everyone able to contribute and comment.
- By being able to openly discuss the ambitions, objectives, problems and contributions of each member of the team, a major step forward is achieved in communication and in bringing about a commitment to the project.

Start-up meetings have several practical benefits and results.

- To bring together the key contributors and to focus on the problems of organization and design which have to be solved.
- To identify the critical work tasks and interfaces that will need the majority of management attention in the future.
- To agree the sub-objectives and their scope, and to secure the team's commitment to their achievement.
- To foster and encourage communication within the team so that the exchange of information can take place smoothly on a regular basis and at the start of each key stage in the project.
- To focus on the need for production efficiency and quality of delivery.

6.5.1 The timing of start-up meetings

The key stages during the design process when start-up meetings should be held are at the start of briefing, scheme design, and engineering design. At each of these points, new people from new organizations are introduced. Usually they will have little knowledge of the project and may well have different priorities. If potential conflicts are not dealt with promptly, productivity may be lost through ambiguity, misunderstanding, and duplication of effort.

The designers should also be involved in the start-up meetings at the beginning of construction and, on large projects, at the beginning of key stages such as: substructure, structure, envelope, plant rooms, and service risers and internal finishes.

6.5.2 The briefing stage

The object of the start-up meeting at the briefing stage is to initiate the preparation of 'the statement of need'. This is the key document required by the project manager to enable him or her to brief the design team properly (see Checklist 2).

The purpose of the meeting is to bring together the key contributors so that a mutual, in-depth understanding of the client's development objectives can be achieved. This includes:

- the client,
- the client's representatives,
- members of each of the client's departments who need to be consulted in the preparation of the brief,
- the client's advisers,
- the briefing team leader.

Checklist 2. Agenda for 'Briefing Stage' start-up meeting

Those to attend:

    client project manager
    client's representative
    briefing team leader
    relevant client specialist department representatives
    consultant design team
    construction manager (if appropriate)
    specialist contractors.

Objective:

    *to develop the statement of need
    *to develop the concept and scheme design
    *to develop the detail design.

Agenda:

1 Introduction – client representative.

2 Why are we here?
    The commercial background to the project.
    The brief as developed so far.
    Round table analysis of the information to ensure that the team fully understands the task.

3 Who are we?
    Brief presentations from each team covering people, skills, roles, responsibilities, their organization's structure and communication systems.
    Workshop to develop communication and organizational map.
    Integration of map by design manager, critical review and agreement.

5 Where are we going?
    Constraint evaluation.
    Potential design solutions.
    Alternate evaluation and progress strategy.

6 Agreement on management of the plan and methods of review to achieve the plan and objectives.

* delete as appropriate

It is intended that the meeting will generate discussion, that ideas will germinate and that an understanding of the values and capabilities of the client and the design team will be achieved.

6.5.3 At the scheme design stage

Once the brief is clarified and the project approved for the next stage, a meeting of the developed and enlarged team is necessary. This will bring together:

- the client's representative,
- the briefing team leader,
- the design consultants,
- the client's functional specialists,
- specialist advisors,
- specialist contractors,
- the construction manager (if appropriate).

The objective is to decide the scope of the individual studies necessary to develop the project to the completion of the scheme design. This is a stage of intense creative activity and evaluation of alternative strategies for the building, requiring a harmony of thought and approach. This meeting is designed to achieve the necessary collaborative working relationships.

6.5.4 At the engineering design stage

Once the scheme design is approved and the design team members appointed, the meeting initiating the engineering stage (see Checklist 3) can be held. It will bring together:

- the client's representative,
- the client's project manager,
- the design consultants,
- the client's functional specialists,
- specialist advisers,
- specialist contractors,
- the construction manager (if appropriate).

The principal aim of this meeting is to achieve a sympathetic working relationship and a co-ordinated approach between the differing styles and cultures of the design consultants, functional specialists, and the designers from specialist contractors. On major projects there may be a need for a start-up meeting for each zone of the building.

Checklist 3.   Agenda for 'Engineering Stage' start-up meeting

---

Those to attend:

  client project manager
  briefing team leader
  client's representative
  construction management (if appropriate)
  design team – ALL design professions
  project control/cost engineering
  the contractor (if appropriate)
  specialist trades contractors (all who have a significant design input. The implications are, either the subcontracts must be purchased, or arrangements have been made to obtain the subcontractors willing participation).

Objective:

  to plan and organize the delivery of the production information.

Agenda:

  1 Introduction – client project manager.

  2 Statement of objectives.

  3 Introduction of participants:
    background
    knowledge base
    intended contribution to the process.

  4 Clarification of perceptions of the individual roles.

  5 The plan generation process:
    statement of project sub-objectives;
    ● definition of zones
    ● definition of interfaces
    analysis of work stages;
    identification of information needs and knowledge inputs.

  6 Plan commitment.

  7 Agreement on management of the plan and methods of review to achieve the plan and objectives.

As the construction information is developed, it is equally important that each consultant understands the others' need for information and when decisions will be required.

6.5.5 The structure of the start-up meeting

Team commitment can be built up by moving progressively from broad strategy to the more detailed implications and tasks. As each member of the team understands the integrated nature of their contribution so the traditional negative barriers will be broken down and a more co-operative approach developed. The meeting should not be overly large or formal and should not be treated as a public relations exercise.

It should be taken seriously and is so important that the participants must be present for the whole meeting, which should take precedence over anything else. The long-term benefits far outweigh any short-term loss of time of the staff attending the meeting.

6.5.6 Agenda for the start-up meeting

An agenda must be prepared for the meeting and should be circulated at least a week in advance. It should include the following.

- *Introduction* – This is generally made by the client's representative or project manager who sets the pattern for the day, outlines the history and background to the project and clearly sets the objectives.
- *Presentations* – Each organization attending the meeting then introduce the key people from their organization and make a formal presentation on the scope of their work and how their contribution will be made.
- *Clarification of roles and responsibilities* – Should there be any misunderstandings about the interpretation of any individual roles and responsibilities these must be clarified by the project manager and mutually agreed at this point.
- *Strategic plan* – The strategic plan should be developed to meet the objectives for the engineering design phase. If this is to be meaningful, the people attending the meeting must come fully prepared and be able to commit themselves and their organization to whatever is agreed during the discussion.
- *Discussion* – This must be focused on the practical steps necessary to meet the next set of objectives and everyone at the meeting must feel free to contribute to the discussion. Where there are barriers to co-operation and contribution, they should be identified and resolved immediately.
- *Conclusion* – The meeting should not finish until everyone is in agreement, understands the plan, is committed to the plan, knows that they can execute the plan.

6.5.7 The design manager at the start-up meeting

The design manager's role is essentially that of organizer and catalyst. To be dogmatic and dictatorial would defeat the objective. The design manager may be from one of the consultant designers or from another professional organization. Architects and engineers have a natural responsibility and authority implicit in their task and may strongly resist the seeming intrusion of a formal start-up meeting into their management of the work. However, experience has shown that an effective session overcomes any resistance by rapidly establishing a positive climate for work. It is through the start-up meetings that widespread commitment to the project is created. This commitment must become infectious and must be communicated to everyone who is remotely concerned with the project. Each project manager, in each organization, should be encouraged to gain the commitment of his or her own team by holding a start-up meeting to bring together that particular team. It is the overall project manager's responsibility to support and promote this course of action.

6.5.8 Arranging the start-up meeting

This should not be an improvised meeting. Each contribution must be properly prepared and carried through with commitment. Many of the people involved will not have encountered this approach before and will probably not be experienced in making even a brief presentation of their task. They should therefore be given training and encouragement so that they perform their task well. To achieve the best, every participant must be: briefed, comfortable in their subject, be reasonably relaxed, and have good presentation material. As start-up meetings can be large, an informal, theatre style presentation is preferable, with facilities such as: overhead project, data projector, and slide projector, as well as video and so on to help avoid monotony.

There should be a pre-meeting briefing session at which those giving the presentation will meet with their own teams to plan the outline of what they will present. The meeting should not take more than 1 hour and should be held sufficiently early to allow enough time for preparation.

Each organization's presentation should include the following.

- An introduction to its team.
- Their interpretation of the project objectives and a clear, concise description of their task as they see it.
- A description of special technology they may intend to use, (e.g. CAD systems or particular design or communication technology).
- How they intend to meet the project objectives in terms of commitment, resources and method.

- Their approach to quality control and zero defects.
- Their policy on designing for health and safety in terms of site work, commissioning, maintenance, occupation and disposal of the building.

Where possible, and if appropriate, samples of relevant technology or models should be used to enhance the 'feel' of the subject. Detailed drawings are not a good presentation medium, as an audience cannot see the detail unless it has been enlarged for use on an overhead projector or slide projector. It is important to emphasize that good content has precedence over a slick presentation. Each presentation should take between 5 and 10 minutes.

6.5.9 Organizing the meeting

For most projects a full day may be required, particularly if many people from each organization are involved. The pressure to compromise on attendance should be resisted. It is important to gain commitment and while the participants may largely view the day as an encroachment on their valuable time, it is vital for them to attend to achieve the development of their own individual objectives into those of the project objectives.

The timetable should be flexibly arranged to achieve:

- by first coffee break, a clarification of individual roles;
- by lunchtime, an understanding of everyone's sub-objectives;
- by the end of the day, a consolidated plan and commitment to it.

Everyone is required to give and share information, perceptions and objectives. As much time as necessary must be allowed to enable individuals to ask questions so that they are confident in the proceedings. If the day is too tightly scheduled, it will not be so successful.

An integral part of the start-up meeting is the plan generation session. The most successful implementation of this idea has centred on a 'planning wall'. This is a wall-size blackboard or white board divided into time periods on which everyone can record their tasks and plans so that each has an opportunity to see what is going on and contribute freely. The project manager records the final, agreed plan and circulates it to everyone. The implication is that everyone who attends can make commitments on behalf of their organization and the organization also stands by any commitments that are made.

7

Defining the tasks

The most important phase is initiation of the project, as decisions made at this point will set the pattern for all subsequent activity. It is essential to provide a clear means for communicating the design and task objectives across the interfaces between one stage and the next. The task is divided into stages. The first three stages are (see Figure 19): getting started, the statement of need, and the business case.

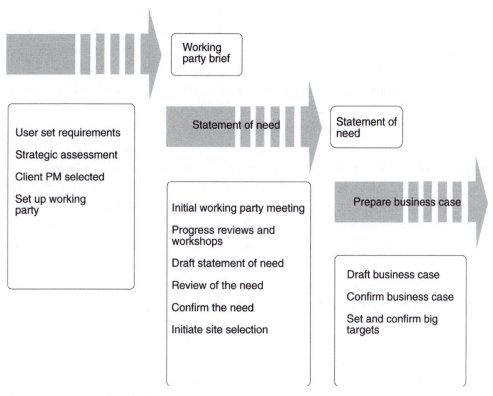

Fig. 19. First stages of the brief generation process.

A simple diagram, which illustrates the basic steps in the process of initiating a project and setting a brief for it cannot embrace all of the permutations necessary on every project. Nor is there a specific time-scale, as this will depend upon the organization initiating the project. It may seem that this is not a stage when a design manager should be involved. However, it is a vital design stage as a very large amount of information is assembled, evaluated and integrated into a set of documents, which are crucial to the success of the project. Whoever is involved at this stage must appreciate the design issues implicit in the process.

7.1 Getting started

Comprehensive briefing is of fundamental importance to the success of the project (Duffy, 1994). However, perfect knowledge is unattainable (Cyert and March, 1963) so it is impossible to establish every need in total detail at the outset of the project. The briefing process is a series of stages of increasing detail as issues and specific requirements are refined and stated. It is important to recognize the need for key decisions at the right time and to stop the iterative process from continuing for too long. Failure in this will lead to dissatisfaction and low morale as the project objectives are continuously questioned and modified. However, modern business cycles are putting this requirement under considerable pressure. Modern business cycles are much shorter than the construction process and business requirements may force significant changes during the life of the project. This leads to potential conflict. The building must be designed with as much flexibility for the users and so the design team is constantly trying to reconcile these needs. This leads to a constant interaction with the client's team to define their needs. If this gets out of hand the project could descend into uncertainty and chaos.

An alternate strategy is to leave the briefing process to as late as possible, compress the design and construction times to the minimum and so meet as closely as possible the real business needs. This approach places emphasis on a briefing process, which when complete, must be totally complete. In practice this leads to the brief being generated for parts of the building in parallel so that it can be linked more closely into both the business process of the client and the project process. The effect of current practice is to place much greater emphasis on getting as many strategic issues resolved as early as possible and then leaving their refinement to a series of iterations when expertise and information is made available as the design process unfolds.

The client and/or the project initiator should establish a working party (see Checklist 4) to oversee the development of the brief. The activities indicated in Checklist 4 should be directed by the initiator of the project. A working party may be set up consisting of interested parties and specialist consultants to assist in establishing the statement of need. The report produced is used as the basis for either confirming or rejecting the need to proceed.

Checklist 4. Action by working party at first meeting

Appoint chairman, usually client representative

Prepare and agree initial statement of need (the initial project objective)

Determine outline requirements of each member

Identify knowledge gaps (using Checklist 6)

Allocate responsibility for determining missing knowledge

Prepare schedule for data collection

Where appropriate, initiate research to establish requirements and specification

Prepare programme and schedule of meetings to achieve the initial review date

Establish success criteria

7.2 Statement of need

The first step is to establish a set of criteria that will form the basis for assessing and establishing the need for the building. These may be:

- the function,
- the timing,
- the priorities.

This is followed by the preparation of a detailed description of the building's intended function. A detailed list that identifies everything to be accommodated both inside and outside the building envelope should be prepared. Checklist 6 is a summary of the basic information required to prepare a statement of need, i.e., to establish the extent and relationship of all the different activities and functions that need to be accommodated in the building.

At this stage, the project should not be thought of in terms of shape, size, or number of stories, but entirely in terms of function. To understand fully the requirements for a building, it is helpful to prepare flow charts of the activities to be accommodated, together with diagrams illustrating the relationships between major areas of activity, core areas, circulation spaces, entrances and items of plant and equipment. The numbers of people involved, the size of the plant and equipment, and the volume of information and materials they will process should all be listed.

In cases where activities need to be kept separate, the reasons should be given. The need for separation may be based on considerations like safety, confidentiality, status, comfort,

Checklist 5.   Action at stage 2 – the initial review meeting

---

Identify any major problems with availability of data
  review Checklist 4 for each interest

Establish criteria against which initial statement of need can be reviewed

Review need to proceed as one of:
  i.   Need established – proceed to functional brief
  ii. Need established – not urgent, action suspended
  iii. Need/Not established – reassess initial objective or suspend action

Confirm the need and establish project priority
  prepare formal statement of need

Submit statement of need to client management for formal review and obtain signed-off brief as outline project approval to proceed to concept and scheme design stage

---

and cleanliness. The acceptable limits of the incompatible activities should be specified. For example, where activities have incompatible noise levels, the acceptable and generated levels should be given. Activities that need to be adjacent must be identified, together with the type of interaction that will take place between them. Those activities that could equally well be housed in one space, or need to be kept separate should also be identified.

It is important to emphasize any unusual or special requirements and for copies of existing technical descriptions of work or its organization to be made available. If possible, it is advisable not to think in terms of any similarities with facilities in existing buildings, because some aspects of building technology change so rapidly that copying the technology of a previous building may not provide the most cost-effective solution.

Where new functions can be accommodated within an existing building or facility, all relevant details of the building, such as size, use and form of construction, should be available. Any legal or physical restrictions on the way that the building can be adapted for future use should be identified, for example, restrictions on use or floor loading. Any known limitations or restrictions on use should be identified with any natural or artificial feature that needs to be retained.

The statement of need should say whether suitable land is available for the proposed development and its location. If the land is not yet in the possession of the client, the proposed method of acquisition and timing should be identified. If available, surveys of the land or any relevant information should also be supplied. Where shared occupancy is envisaged, any constraints imposed by current use, such as restrictions on dust levels, noise/vibrations and sensitive operations should be classified.

Checklist 6. Information for the statement of need

1 Activities and functions (for each activity give the following information):
  - a thorough description of the function or use
  - the number of rooms and/or open areas required
  - additional access, area for maintenance of equipment
  - the numbers of people involved
  - the space required for each activity
  - the required minimum floor-to-ceiling height of each space
  - frequency of use – continuous, permanent, or intermittent in hours or days
  - special equipment, function, type, size, weight.

2 Services required by each function:
  - telecommunications – BT, wireless and satellite communications
  - air conditioning
  - ventilation, extract
  - lighting
  - power – low voltage, high voltage, vacuum
  - piped gases
  - effluent treatment
  - wastes
  - water – potable, treated, H & C
  - heating/steam
  - IT
  - fibre optic links.

3 Environmental control:
  - temperature, air changes, pressure zones, dust/particle control, fume extraction, bacteria control.

4 Cleanliness standards.

5 Environmental conditions.

6 Security:
  - privacy
  - soundproofing
  - safety – theft, access, egress, fire.

7 Hazardous activities:
  - radiation
  - pollution
  - chemicals
  - bacteria
  - noise and vibration.

8 Wall, floor, ceiling finishes:
  - specify standards and specific type of finishes required.

9 Relationships:
  - relationships with which room/activity/function
  - define if room/activity/function is associated.

10 Interrelationship between function and local environment

11 Facilities provided by other related facilities:
  - services – type, quantity, volume
  - waste treatment
  - storage
  - component/material supplies
  - computing, data
  - people movement
  - access, egress, size, shape, flow rates.

12 Environmental factors:
  - impact on landscape
  - security
  - visual appearance
  - access
  - delivery vehicles – size, shape, frequency, time
  - parking – permanent, visitors
  - protection against pollution of the environment with scale and type of pollutants
  - planning restrictions – constraints, public enquiry, conservation areas, listed buildings, preservation orders, rights of light
  - requirements of statutory authorities and services companies
  - site surveys of available sites.

The statement of need should identify time constraints relating to the vacation of existing accommodation, to current or proposed land occupancy, or to any other deadlines and key dates which must be met. It should also consider how long it would be before the various activities and functions need to be changed to meet future anticipated requirements. Such information will enable a decision to be made on whether it is likely to be more cost-effective in the long term to design the building with built-in flexibility.

The working party should then examine the statement of need from every point of view to establish whether or not a new building is necessary or financially viable. During this process many alternatives will be considered and options developed. Each should be evaluated to ensure that the best arrangements for the client are developed (see Checklist 7).

Checklist 7.   Questions to ask at the options stage

1 How appropriate are the existing facilities for the client's needs?

2 What proportion of the existing facilities are utilized and could be improved?

3 What is the current condition of the existing facilities and how could they be improved to meet the requirements?

4 How do the standards of the client's activities compare with best practice?

5 How does the way that the organization runs and manages its assets compare with best practice?

6 What resources are available to the client for the improvement either of the existing, or for the construction of a new, facility?

7 What are the costs and benefits of each of the options identified?

8 What are the key constraints and risks associated with each of the options identified?

The working party must also assess the level of risk to be incurred, based on the accuracy of the information available. To ensure objective analysis and evaluation it is necessary to establish criteria, such as return on capital investment, against which the statement of need should be compared. Once the need is established, a formal statement of need can be prepared and presented to the client for approval.

7.3 Business case

The agreed statement of need should then be developed into a sound business case. This should consider the impact on the business both physically and financially. Financial resources will be required and how they are to be provided must be considered. In many cases this document is no more than one page to be presented to the board of the client for consideration. Many clients have established procedures for this proposal, particularly if they have a large capital expenditure programme. This is a crucial stage in the project because once this stage is passed the budgets and priorities of the project are largely set.

Clients may try to use this stage to bring about a change in the way that the construction industry operates. They are not content with past performance or become aware that many clients are obtaining far better performance from the industry on their projects. Many clients use sophisticated business process reengineering of their business and seek similar responses from their construction operations. The broad economic environment for construction has been affected by the demands set out by Latham (1994) for the industry to reduce costs by 30% by the year 2000. This has been followed by Rethinking Construction, in which Egan (1998) has set a series of continuous improvement targets for the industry. Clients are, therefore, benchmarking their requirements against the performance of others and setting very challenging targets when they approve the business case. This has major effects on the subsequent stages.

Inevitably there needs to be some flexibility in the targets that are set at this stage. The development of the functional brief, its translation into the concepts and scheme design may mean that the original ideas will change. However, if the risks have been assessed thoroughly, and a contingency set against the risks, then this can be managed through a responsible reporting procedure to the client.

7.4 Functional brief

At this stage of the project the business objectives and the statement of need are translated into a researched and comprehensive functional brief, which precisely defines the client's requirements. From this, a clear definition of the project and reliable budgets of cost and time can be produced for final approval.

For smaller, simpler and more straightforward projects, the original group that developed the statement of need can continue to develop the functional brief (see Checklist 8). On larger and more complex projects there will be a need for a significant development in the skills of the project team. In projects where budgets and targets are particularly demanding or set at the leading edge of practice, there is a trend for a multi-disciplinary team to be established. Depending on the project, the team may include:

- in-house functional experts,
- an architectural team,
- a structural engineering team,
- a services engineering team,
- specialist consultants, e.g. planning legislation, acoustics, landscaping, process planning,
- a cost management and cost engineering team,
- project and construction managers including: constructors, production engineers, design managers, and value managers.

Current experience of leading practice shows that these teams need to draw upon the most experienced people from each of the contributing organizations. This stage involves a large degree of conceptual analysis that draws upon past experience. Therefore, the group needs to have a large experience base. Where this is not available, analytical methods such as value management and quality function deployment, can be used to structure the development of the functional brief.

Checklist 8. The functional brief

The functional brief will include:

1 Descriptions of:
- the client
- the functions or uses to be accommodated (room data sheet format)
- any requirement for future flexibility or expansion
- required services
- functional quality standards
- standard brief, modified to the specific requirements
- statutory/safety requirements
- cost and time estimates
- prioritization of objectives
- value analysis, buildability and hazardous operations studies.

2 Outline plans of the proposed building which show the arrangement of spaces and their use.

3 A budget or cost estimate which should include separate items for:
- construction
- commissioning
- design fees
- site management and organization
- inspection and testing
- furniture and equipment.

4 Client's direct and indirect costs of:
- moving, and where appropriate, temporary accommodation
- hardware
- operation and running costs
- maintenance
- cleaning and servicing
- marketing.

The functional briefing stage is an active time for all concerned. The client will be required to furnish further detailed information and to make decisions about the acceptability of possible design solutions. It will be necessary to advise and make decisions on matters such as:

- limitations imposed by existing buildings and services,
- land acquisition,

- finance,
- approvals procedure/system,
- consultation periods,
- workforce management.

The development of a well-researched, comprehensive, functional brief, which truly represents the client's needs, is acknowledged by all in the industry to be the key to success. Therefore, it is imperative to allow adequate time and funding for a full briefing process and to appoint, where appropriate, a client's representative of the right calibre, experience and seniority (level IV or above, see Chapter 5). The client's representative needs to be supported by an appropriate organization and expertise.

Once the functional brief has been established, estimated and agreed, changes should be resisted at all costs, although the practicalities of this were recognized at the start of this Chapter. Apart from the potentially excessive additional expense associated with abortive work, 'knock-on' effects, delays to programme etc., fundamental changes made to the scope part way through the project compound the inherent difficulties of the building process. The use of value management or the discipline of quality function deployment techniques may help to control the need to revisit the earlier stages.

7.4.1 Value management

Every client wishes to achieve value in return for the investment. The specific criteria of value, however, depend upon the particular client. For successful projects these criteria need to be established, hence the need for value management (VM).

> 'Value Management is defined as: a structured approach to defining what value means to a client in meeting a perceived need by establishing a clear consensus about the project objectives and how they can be achieved' (Connaughton and Green, 1996).

Value management is a strategy of examining every aspect of the whole project to ensure that all of the expectations can be delivered in the most economical way. Often the criteria are broadened to embrace the operational and lifetime costs of running the facility. The introduction of value management is via a value management workshop either at the functional brief stage (concept development) called VM1, or at the feasibility stage (planning permission) called VM2, or at both. The characteristics of the project and the client organization will determine the most appropriate approach. A more complete description of value management is provided by Connaughton and Green (1996).

The kind of approach offered by techniques such as value management provide analytical tools and recording methodologies that are very important in helping clients and designers to articulate and prioritize their ideas. Good briefing is notoriously difficult to achieve because the usual format, the written word supplemented with drawings, is open to wide interpretation, particularly at later stages of the project. As a project develops, the number of people involved changes rapidly, with many arrivals and departures. Each new person will have a different interpretation of the requirements, partly because they have not been quantified or established against a recognizable base. The development of analytical techniques enables the brief to be used throughout the project as the basis of all judgements on the relative values of the issues.

A problem with traditional briefing documents is that few contain any judgement as to the relative priorities in the requirements. The assumption is that all needs must be totally satisfied irrespective of the costs or value attached to them. However, many developer clients are questioning this assumption. For example, is it necessary for office fire escapes to be finished to the same standard as normal internal stairs? There is a strong architectural school of thought that they should be, since every part of a building represents the design ethic. However, a balance between the respective contributions to the utility and value may enable money to be switched between areas to give better overall value. Stating these as variable requirements, each with a level of acceptability at the beginning of a project, requires a different structure to the briefing process and its documentation.

● *Value hierarchy* – The value hierarchy is a method of taking the primary project objectives and breaking them down into their sub-objectives. Each sub-objective is the means to achieve the main objective. The sub-objectives can be further subdivided (see Figure 20). This process can go through several iterations as part of

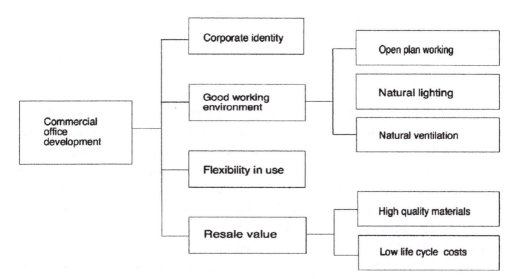

Fig. 20. VM1 value hierarchy.

VM1 and VM2 until the team and client agree that this is an accurate statement of the project's objectives. This statement can either stand alone or be developed further.

● *Value tree* – A process of weighting the objectives and sub-objectives of the value hierarchy produces an ordering of priorities between all of the conflicting demands in the project (see Figure 21). The weights are arrived at by group consensus during the workshop.

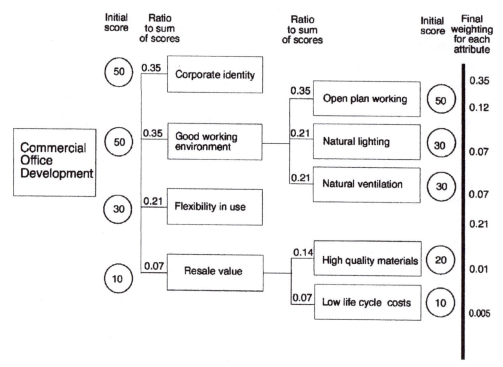

Fig. 21. Weighting of the value tree.

● *Decision matrix* – Once the weights to the various project objectives and sub-objectives have been set then it is necessary to decide which of the options will provide the best value when set against the objectives, i.e., how well they meet the targets that have been set (see Figure 22). Many of these assessments are subjective, but the essential characteristic is that the decisions are a consensus view of everyone involved in the process. The quality of the assessment will depend on first, how well the initial objectives were defined, and second, on the expertise of the project members. The decision matrix is a very good tool for use during the options analysis stage as well as the later stages during the initial developments of the concept design.

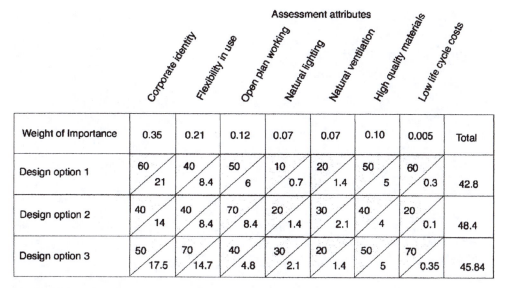

Fig. 22. The VM decision matrix.

7.4.2 Quality Function Deployment (QFD)

A technique that has many similarities to the value tree and decision matrix is QFD or the 'House of Quality'. Originally developed within Japanese manufacturing industries, it has been used to achieve considerable improvements in product design. It has recently been developed as a tool for use in construction projects (see Figure 23).

At first sight, the tool is complex in that it has six parts. However, it only requires an understanding of the principles of matrix analysis with the addition of quality and benchmark criteria (see Figure 24). The other advantage is that QFD carries forward, through the key stages of design development, the initial project values and criteria in a way that ensures everyone is consistently working to the same values. In this way there is little dispute, reinvention or interpretation of the initial client values at any stage in the project, thus increasing the efficiency of the design process. It is claimed that QFD can halve the time and effort of the design process because it enables consistency and the removal of ambiguity. The 'rooms' in the house of quality are as follows.

- *Room 1: The primary objective* – The client must state clearly the primary objective of the project. This needs to be a brief statement that includes the key contribution that this project is intended to achieve.
- *Room 2: The whats* – The client's objectives are subdivided into ways that the overall objective can be satisfied in the same way that the value hierarchy is broken down. The requirement is that all of the issues, including those not normally stated such as robust structure, are also stated such that there are no areas missing. Each requirement in the list is then ranked for its importance.

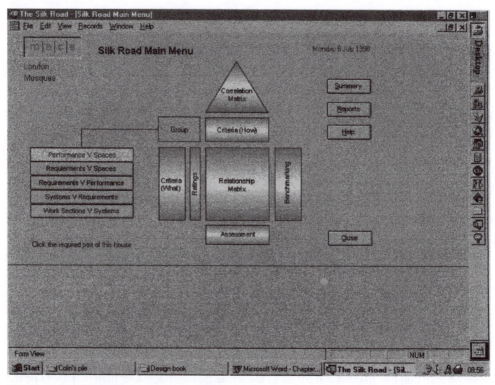

Fig. 23. QFD software for the briefing process.

Requirements	Importance rating	Comparative project analysis					Rating				
				Project			Highest score	Poor		Good	Highest
Project Z	Project Z	A	B	C	D			1	2	3	4	5
What No 1	5	2	1	2	4	4				
What No 2	2	1	1	2	1	2				
What No 3	1	3	3	2	1	3				
What No 4	3	1	1	2	2	2				
What No 5	4	5	4	4	5	5				

Fig. 24. A benchmarking approach within QFD.

Quality of living for pre-fabricated housing	Design objects	Indoor environment									Projects Analysis			
		Openness	Cross ventilation	Draftiness	Heat insulation	Heat retention	Condensation prevention	Noise insulation	Floor insulation	Quietness of appliances	Project A Rating	Project B Rating	Project C Rating	Target Level
Project requirements / IR														
Good indoor environment — Fresh air / 4		9	9	9							3	5	4	4
Proper humidity / 5		9	9	9	1	3	3				4	4	5	5
Proper room temperature / 5		3	3	3	9	9	1				3	3	4	4
No leak of indoor sound / 2		3	1	1	3	3	3	9	3	9	3	4	4	3
No noise from outside / 3		9	3	3	1	1	1	9	1		3	4	5	3
No vibration from outside / 1		1	1	1				9	3		2	4	3	2
Absolute Scores		130	108	108	59	69	29	54	12	18	587			
Relative Scores		22%	18%	18%	10%	12%	5%	9%	2%	3%	100%			
Competitive Scores		131	107	107	53	63	31	72	18	27	609			
Competitive Assessment		22%	18%	18%	9%	10%	5%	12%	3%	4%	100%			

Fig. 25. A completed QFD matrix of a design element.

- *Room 3: The hows* – A list of all of the ways of satisfying the whats is recorded across the top of the matrix.
- *Room 4: The relationship matrix* – All of the hows are assessed against each of the whats on a sliding scale: 0 for no link, 3 for a partial satisfaction, and 9 for a strong satisfaction.
- *Room 5: Technical assessment* – The scores for the hows are totalled, from which the most satisfactory method of solving the particular requirement can be seen. This can be given as an absolute score or a relative score.
- *Room 6: Competitor analysis* – Quality is a subjective issue. Building design is more difficult than product design because a building cannot be put on the table, analysed and dissected like a consumer product. However, this section of the QFD approach helps overcome this basic limitation of a normal brief, where the quality is not quantified and is delivered largely through a faith in the lead architect and key designers' ability based on past performance. QFD requires that the importance rating be set against an assessment of competitors' performance. This statement first requires that competitors' products are assessed, and secondly, a decision made as to whether the new product is to meet or exceed competitors' products. Naturally this requires considerable research

and evaluation, but this is the secret of this technique because every aspect of the new product has a measurable and visual goal. To apply this to buildings requires a similar research and evaluation of previously constructed buildings and a decision as to whether the new building is to be built to the same standards and quality or to a different standard.

Rooms 1–5 constitute a basic approach to the use of QFD. Room 6 involves benchmarking against external organizations, which requires significant data collection and analysis. This may be uneconomic for a one-off project.

Not everything can be dealt with in one matrix, particularly something as complex as all the requirements for a building. In practice, the QFD approach can operate at three levels within the briefing process. At the highest level it could set the detail for the functional brief. The systems that would satisfy the hows would form the whats for the next stage of system design during the development of the concept and scheme. The final stage would be detail design where the subsystems that satisfy the hows become the whats of the fine level solutions. Because of the consistency of the method, which links at all levels back to the satisfaction of the client's real priorities, this is a very powerful way of recording the briefing process and the value judgements made within the process. Again, the judgements have to be made through consensus of all involved in the discussions.

To help in the process of deciding and selecting the most profitable ways of solving the whats, a correlation matrix is established that assesses the compatibility of the hows with each other. It is no good selecting a series of solutions from the hows if they will not work with one another. The correlation matrix is a simple approach to help avoid such incompatibilities.

By quantifying the basic briefing decisions the client and the whole team are aware of all of the implications in cost, production, and final quality of every aspect of the building from the beginning.

7.4.3 Standard briefing documents

The purpose of a standard briefing document is to avoid 're-inventing the wheel' for every aspect of the design. It is particularly suitable for clients who have a continuous development programme and who can benefit from carrying forward the experience gained on one project to the next.

It is not intended to be a straightjacket for design and the commonly used documents are usually confined to the application of construction technology that has been refined over many projects. The advantage of this approach is that it avoids discarding valuable experience from earlier projects.

Similarly, when new designers are commissioned, they can be rapidly briefed on the forms of construction and the lessons learned in previous buildings. The document can

Checklist 9. Typical standard brief provision

Precast concrete construction

A precast concrete frame that incorporates the structure, floors and external cladding takes advantage of the total benefits of the system. A dry construction, watertight envelope can be achieved more quickly than with other forms of construction.

- Get to understand the range, flexibility and cost before choosing the system.
- Use standard elements wherever possible; if necessary, change the applied finishes or adjacent building elements, not the precast units.
- Do not extrapolate, particularly structurally, without due care, understanding and experience.
- Embrace the system and make the most of it rather than battle against it.
- Under certain conditions a precast frame is 14% cheaper than a steel frame and 21% cheaper than a composite floor design structure.
- Use the maximum spans available in the system.
- Minimize the variability of frame sizes and sections.
- Use spandrel panels to incorporate the edge beam and cladding into one component.
- Precast floors are manufactured on high quality factory beds and may be suitable, in many cases, for direct decoration without the need for expensive further treatments.
- Avoid the use of structural topping screeds. Use tapered edges and in situ concrete fill. Allow sufficient space for the concrete to be placed into the joint and for compaction.
- Avoid structural links to steel frames that require welding onto the steel members during manufacture.
- Ensure that the erection can proceed in a regular manner without needing to use overly complex arrangements of equipment.

Source: Guidelines for the use of precast concrete frames, The British Precast Concrete Federation

---

also contain experience and guidance gathered from recognized authorities in the industry (see example in Checklist 9). In order not to restrict innovation, the ideal standard brief would contain a challenge to produce even more efficient design solutions that may, if proven in practice, form the basis for the next issue of the document.

Typically, alternative designs are evaluated on the following criteria:

- value,
- improved flexibility and performance of the building,
- reduced costs of occupation,
- improved speed of design or construction,
- better quality or more reliability.

It is good practice to bind the statement of need together with the standard brief, suitably modified for the particular project, into all agreements and contracts for the project.

As the functional brief is the culmination of the briefing process it is important that it covers all of the client's requirements and is totally accepted and agreed, and formally signed off by the client and the design team.

7.5 Concept and scheme design

The development of the functional brief into a full scheme design is a substantial exercise that must be done carefully in conjunction with the client. Sufficient time must be allowed, particularly if the project is complex, for the large amount of investigation and research necessary to investigate fully, value engineer, cost, programme and finally agree on, the principal systems in the building. It is clearly very difficult to resolve every issue simultaneously. However, for efficiency, once the basic concept and its feasibility are accepted, the various system designs can be developed in parallel, as long as the technical and physical interfaces between them are isolated and the boundaries defined (see Checklist 10).

Where this is not done, either the design will become a protracted sequence of interlinked decisions, or work will continue, but will have to be repeated when the detail of the interdependencies is resolved.

The completion of the scheme design is a crucial client decision point in a project. It is also a key point in the management of the process. The client and the designers must agree on the complete scope of the work. It is usual at this stage for the design consultants to prepare a complete set of project information that is formally agreed by the client and will include the following in the detail listed in Checklist 9.

- plans
- elevations
- specification
- cost plan
- programme
- special studies and reports.

Once the scheme design has been reviewed and agreement has been achieved, the complete document set is signed off by the client and design team, and the next stage can begin on a basis of certainty.

7.6 Engineering design

Engineering the project in this context is defined as the preparation of the details necessary for site production. It is the turning of the design, formally agreed at the end of scheme

Checklist 10. The scheme design

1 Architect:
- floor plans showing the details of all spaces and their use
- details of main cores and risers
- sections through the building showing the key relationships between spaces
- elevations of all facades showing the characteristics and relationships of all components
- perspectives and details of key architectural features
- site layout
- main services layouts
- existing site surveys
- finishes schedule for all spaces.

2 Engineering:
- floor plans showing layout and sizes of all structural components
- key sections through the structure of the building
- specific details of all non-standard structural components
- floor plans showing all services systems
- schedules and specification of all services systems and key components
- the principles of foundation design
- geotechnical survey
- acoustic treatment.

3 Mechanical services:
- the principles of the heating, ventilation, cooling and plumbing systems together with critical dimensions within ceiling and floor voids
- the principles of the fire protection and fire alarms systems
- plant room locations, sizes and weights of principle items of plant
- location and size of principle service ducts and risers.

4 Electrical services:
- the principles of lighting and power distribution; tenant and client supplies, power and lighting to common areas
- stand by electrical generation
- building management systems
- lighting protection
- security systems
- electrical intake and transformers with size and position of switchrooms.

5 Vertical transportation:
- passenger use studies
- principles and location of alternative systems, and location and size of lift motor rooms.

6 Briefs:
  - developed functional brief into a full brief of the client's requirements.

7 Cost and time budgets:
  - detailed cost plan
  - overall project programme
  - programme of components on long lead delivery
  - outline construction programme.

8 Method statement:
  - a statement of the way the building will be constructed.

9 Approvals:
  - a copy of planning permissions and approvals that have been granted to date or a statement of the position regarding the obtainment of the necessary approvals.

design, into the working information. It should be a distinct phase of the design process beginning once the scheme design has been approved. A clear organizational plan for the engineering stage of the design is required which should be based on the division of the project into zones, each of which is a set of complete and integrated systems, within the building. The reason for this is that for the completion of the site operations, each system must function properly. This will be very difficult to achieve if the responsibility for providing complete and co-ordinated information is not system-focused and clearly defined.

By concentrating on achieving completion of each system, a clear objective is set and a discipline which focuses all contributions on to the operational functioning of the system is established. The lead production designer for each system can then be obliged to ensure that all the information to enable the system to be completed is available at the right time. Without such a clear strategy, responsibility can become blurred, and intermediate activity can become the focus instead of the real task of completion.

In an office building the primary zones are:

- substructure
- structure
- external envelope
- cores, risers, lifts, plant rooms and services, including finishes
- horizontal services distribution and finishes.

The design specializations involved and the natural leader flow logically from these zonal groupings (see Table 8). Who actually takes the lead is a matter for negotiation, but it must be made clear who has the lead in each particular zone. This approach, whilst ideal for the

Table 8. The changing relationships within a zoned design process

Building zones	Lead designer	Secondary designers
Substructure	Structural Engineer	Architect Services Engineer Specialist Contractors
Structure	Structural Engineer	Architect Services Engineer Specialist Contractors
Envelope	Architect	Structural Engineer Services Engineer Specialist Contractors
Cores/Risers	Services Engineer	Structural Engineer Architect Specialist Contractors
Finishes	Architect	Services Engineer Interior Designer Specialist Contractors

supply of construction information, requires accurate policy decisions in terms of definition and co-ordination of the interfaces within zones and between them. As with the 'wheel of dominance' the lead of the whole process at any one time can change in emphasis, and the organization and people within it should have a flexible attitude to their leadership role.

Each zone design leader should establish and clearly express a policy for the following.

● The scope of the production information to be produced by the zone team.
● The production information required to be produced by the specialists and works contractors.
● The method of co-ordination to ensure the information for construction is complete and workable.
● The information for tendering.

There should be a specific checklist for each package within each building zone which expresses each of these policy objectives (see Checklist 11 for a guide to the scope).

7.7 Control of interfaces

If the organization strategy based on zones is adopted, then a clear separation of the systems that cross zones is required, otherwise the interference between them will be such that it will destroy the advantages of the strategy. The interfaces relate to the physical features of the building. They occur between zones, systems, components and locations.

Checklist 11.   Engineering design

The checklists are based on the primary zones in the building.

1   Site plan and infrastructure:

- site layout – measured, line, level and elevation survey
- site boundaries, rights of way, orientation, road names, adjoining buildings and owners, adjoining basements
- existing building and sight lines, rights of light, (in London St Paul's cones and grid)
- structures to be preserved – listed building, conservation areas
- extent of archaeological interest and proposed examination of site
- outline of new buildings with main dimensions and information required to position building on site
- new access and exit points from roads for car parking, servicing including loading/unloading and rubbish disposal, visitors' entrance, new roads and footpaths with widths, levels, falls, details of crossovers, crash barriers, controlled crossings, etc.
- details of temporary and permanent fences and walls on boundaries and within the site
- hard and soft landscaping – steps or changes of level, indication of proposed banking, cutting or other excavation and areas including levels for depositing, storing and spreading surplus soil, tree planting, irrigation, mowing margins, street furniture, flagpoles
- survey of existing above and below ground services
- proposed soil and surface water drain runs, gullies, grease-traps, manholes, roding eyes, petrol interceptors, connections to existing sewers (manhole sizes, cover and invert levels may be shown on a separate schedule)
- position and inverts of gas, water, electricity, British Telecom, Mercury and other services with positions of connections to existing mains or meters – termination points within buildings co-ordinated with design of internal services
- external lighting, illuminated signs, earthing plates, CCTV masts, etc.

2   Substructure:

- soil reports and foundation design assumptions of ground bearing capacity, ground heave, water table, tidal water, artesian pressure, flow rates
- demolition of existing structures, provide a method statement for safe demolition, consider ground heave
- reuse or removal of existing foundations
- existing and reduced ground levels, establish new building datum level
- basements; retaining walls, methods for temporary and permanent anchorages, tanking, waterproofing, provide a preliminary construction method statement, finishes to in situ structure, internal drainage, pumping system
- services undercroft, crawl spaces, access/ventilation, service duct entries

- foundation types, e.g. piling, mass concrete fill, strip foundations
- alternative types of piling design
- excavation, reinforcement and concrete quantities, method statement for materials handling and construction logistics, preliminary temporary works designs, upholding existing pavements, services, basements and structures, assess risk of settlement, construction access and egress from site, safety proposals
- use of basement, location of plant rooms, wind bracing, car parking spaces and layout, preferred column grid, compatibility with main building grid, assess need for transfer structure or special members
- proximity to deep underground structures, tube lines, post office tunnels or trunk water mains
- electricity board's switchrooms, transformers
- lightning conductor and lift pits, thickening to ground slab for drainage and service ducts
- plant support steelwork, plinths, holding down bolts, cast-in fixings
- temporary and permanent access for plant installation, removal and maintenance, smoke ventilation, duct system and effect at street level
- parking lifts and turntables; compactors, rubbish removal
- stand-by generators, oil storage and fill points
- sprinkler systems, water storage.

3   Structure – sections and details of:
- frames, suspended floors including framing and trimmers around openings
- framing plans for steel or precast/in situ concrete frame, dimensioned building grid, offsets, and member sizes
- galleries and mezzanines
- decking systems – in situ, precast concrete, metal decking, composite construction, etc.
- edges and special details
- expansion joints, movement joints, bearing pads, isolating structures and construction
- cast in fixings, components, column protection, service penetrations through structural members
- service risers, shafts and flues
- staircases and ramps
- escalators – fixings and motor rooms
- typical and detailed core layouts
- rainwater collection and disposal system.

4   External envelope
4.1 External walls:
- system performance specification, components, materials for glazed and solid areas, acoustic and thermal performance, special systems
- elevations and sections, panel sizes, grid modules, depth of reveals, mullion centres (co-ordinate with partition layouts)

- ground floor entrances, plans, elevations, sections and details
- separate detailing for all exterior wall variations, systems and junctions, copings, parapets, balustrades, planting boxes
- detailing of special features, expansion and movement joints
- rainwater collection and disposal, lightning conductors, external lighting, signs, sculpture, logos, CCTV and security systems
- coursing rods
- lintel and cill schedules
- window and external door schedules
- waterproofing and damp course details, rainwater collection and disposal
- fixing and restraining details
- window cleaning access and safety harness fixing systems
- fire rating of wall
- opening lights for smoke ventilation
- external sun shading – internal blinds
- air intake louvres
- incorporation of perimeter heating systems.

4.2   Roofs:
- specification of materials with required performance and guarantees
- members, plates, coverings and access hatches
- roof finishes, upstand, weathering and waterproofing details
- walkways, fire exits and escapes, access routes to plant areas and plant rooms, handrails, gates, and escape stairs
- isolated plant and equipment on roof – supports and waterproofing.

4.3   Plant and lift motor rooms:
- elevation and roof covering systems
- ventilators, grilles, access doors, flashings
- access arrangements for plant installation, maintenance and replacement
- plinth and fixing details, strengthening roof slab, access ladders, co-ordination of shaft and service riser positions.

4.4   Atria roofs:
- plans, elevations and sections of all support and structural systems
- glazing details including flashings and waterproofing
- smoke extract systems, controls, power supplies, alarm systems
- patent glazing systems
- cleaning systems.

4.5   Cleaning equipment:
- cradles, runways, garages, holding down systems, power supplies, operative access.

5   Risers, plant rooms and primary services

5.1   Risers:

- air flow and return
- air-intake for air conditioning
- water flow – mains, potable and cold water supplies
- hot water flow and return – heating systems
- low pressure hot water, flow and return, local heating
- hose reels, fire mains, dry risers
- treated water flow
- natural gas supplies
- boiler and generator flues
- high and low voltage cables
- electrical bus bars
- lightning conductor
- telephone cables
- building management systems.

5.2   Soil drainage:

- waste drainage
- rainwater stacks.

5.3   Plant rooms:

- lift motor rooms
- layout of major items of plant and equipment, fuel and water storage tanks including plinths, holding down bolts, drainage, power supply and control cabling routes, oil and water retention bunds, water proofing etc.
- power supplies – switch gear
- controls – panels, cabinets and wiring.

5.4   Air circulation plant rooms:

- air treatment
- heat recovery
- duct work layouts – access points, filters, fire alarms, dampers, insulation
- fans – location and weight, plant bases, power supplies and controls
- external louvres
- mixer chambers
- air conditioning plant
- calorifier
- hot water – flow and return
- chiller
- water supply

- air supply and return
- duct work
- electrical supply
- pumps, fans and controls.

5.5   Low pressure hot water plant rooms:
- boiler, type, flues – stack or balanced
- alternative forms of energy supply – gas, oil, electricity
- water supply
- calorifier
- pumps, controls, power supply, meters
- plant bases.

5.6   Cold water:
- supply, tank storage, expansion tanks, pumps
- potable supply
- toilets and cleaners' cupboards, sanitary ware, wastes, overflows
- meters
- drinking fountains, vending machines
- pumps, controls, power supply, meters.

5.7   Electrical services:
- electrical mains – inlets
- substations – local electricity board requirements
- transformers
- distribution system
- meters
- emergency equipment – batteries, standby generators including fuel systems.

5.8   Finishes to cores and landlord areas:
- finishes schedule
- door schedule
- floor, wall and ceiling layouts – access panels
- lift architraves and finishes
- stair balustrades and handrails
- emergency lighting
- entry phones and video systems
- postal delivery collection
- power points for cleaning and maintenance
- entrance desk – control equipment, building management and energy monitoring system, loudspeaker system, emergency evacuation system
- planting

- entrance doors and mat wells
- special finishes and features including planting, water features, and special lighting effects
- hose reels and fire-extinguishers.

5.9 Kitchens:

- services, ventilation and wastes to appliances as necessary.

5.10 Sanitary ware:

- layout and specification for showers, hand basins, sinks, WCs, urinals, dispensers, mirrors, incinerators and flues.

6   Horizontal distribution and fitting-out

6.1 Services:

- air quantities and circulation
- supply ducts, ceiling/floor duct layouts, outlet grilles – type and positions, temperature and volume controls, space divisions, dampers, etc.
- location of support systems
- extract ducts/ceiling plenums
- smoke extract system – interlinks with air extract
- louvres to atrium
- ceiling compartmentation – smoke curtains
- hot water flow and return and condensate pipe work
- electrical supply and controls
- fans.

6.2 Low pressure hot water heating:

- flow and return pipe work in risers, insulation
- flow and return distribution
- radiators – position, zone controls.

6.3 Access panels:

- walls, ceilings, floors – internal, external
- security installation
- smoke detectors
- alarm bells
- burglar detector and alarm system
- automatic fire-extinguishers.

6.4 Finishes:

- finishes schedule
- internal door schedule – location, numbering system for doors and locks, fire rating, finish, decoration, door swings, door sets or independent frame sizes, architraves, etc.
- cill heights, lintel heights and type

- fire stops to voids
- suspended ceiling, including access to voids and services controls
- ironmongery schedule, master key system
- floor finishes, raised access floors, screeds for solid finishes, change in thickness, concrete floor sealing
- ceiling finishes, co-ordinated reflected ceiling plans – ventilation grilles, light fittings, access panels, sprinklers, loudspeakers, etc.
- acoustic and other special finishes
- layout of small power, phone, and information cabling and outlet points
- internal wall finishes, plaster stops, panelling, decorations.

However, to achieve the design, designers and their organizations must recognize the interfaces and formalize their approach accordingly.

The lead production designer of each zone will have the responsibility to determine the interface policy for every aspect of the work in the zone. Where there is an interface with another zone, there will be negotiation with the other lead production designer to determine the details of the interface. The object of the negotiation is to establish:

- where the line of the interface is to be drawn,
- who has primacy in co-ordinating the design,
- the information requirements of the parties to the interface,
- the physical termination at the interface on site,
- the fixing provisions at component interfaces,
- the policy on tolerances at the interface.

For good buildability the fixings for one component or system should not be provided by another component or system. The tolerances between components must reflect the practical tolerances of the two materials and the practicality of the method of assembly.

7.8 Complete information fit for purpose

Information should only be issued for construction when it is complete, which presumes that it is possible to determine completion. The designer's view of completion is often different from that of the contractor and it is necessary to establish an agreed status for the information, to avoid disputes.

Often a designer will have difficulty in determining the detailed requirements of the site. Knowledge from custom and practice is often inadequate, as the scale of information generation and clarification during construction testifies. Equally, the contractor needs to ensure that designers understand the needs of the procurement process and site operations and that these are reasonable given the normal progression of the design process.

It is important to ensure that the person to whom information is to be given has prepared an adequate description of their requirements so that the designer knows what information is expected. There must also be an understanding by the designer of the level of discretion at the point of use for the interpretation of the information.

Construction using manufactured products, for various reasons, allows very little interpretation of the information on site. Therefore, the lead designer for the zone has to ensure that every aspect of the work is detailed fully and correctly. There is little scope for error in the designer's work. Thus it is vitally important that the designer is aware of the information demand and that the drawing, when issued, must be perfect. This degree of precision is difficult to achieve, unless there has been a clear definition of the information need, and the appropriate resources and time have been allocated to its production. It is important that there is a clear statement of the detailed objectives agreed between each designer and the recipient of the information. An effective start-up meeting at the beginning of the engineering stage, which involves all of the designers, specialists and the contractor could, as one of its objectives, set out to achieve a significant clarification of each party's information needs.

There is a duty on everyone to produce correct and accurate drawn information, but before it is issued for construction there must be one final check. It should be done by the most senior or experienced designer on the project.

The following questions need to be considered.

- Is the drawing needed? Can the information be shown in a better way or is it duplicating other information already given?
- Are the dimensions complete?
- Are the tolerances practical? Are they in accordance with the relevant British Standard, or a previously agreed project policy, or are they being set as a preliminary basis for negotiation? The first two are realistic but the third approach may lead to considerable site problems if the targets are unreasonable.
- Are materials and joints compatible? Is the joint system in accordance with best practice and does it avoid cathodic action between materials?
- Are the fixings within an agreed range of sizes to minimize variation?
- Are the specification notes on the drawing complete and accurate and adequately cross-referenced?
- Has a buildability analysis been completed (see page 128)?

Once the check has been completed the drawing should be stamped 'Approved for Construction', assembled into the relevant set and issued.

8

Managing information production

'It has long been appreciated that when the information provided to contractors is insufficient, conflicting or incorrect, this leads to problems on site with a consequent reduction in the quality of the work, delays and increased costs' (Gordon, 1987).

This statement introduces a report of a multi-disciplinary initiative to improve the planning and production of design information. Called the Co-ordinating Committee for Project Information (CCPI), the aim was to produce guidelines for a consistent approach to documentation of design. To this end, they produced codes of procedure for specifications, production drawings, and bills of quantity. The production drawings code of procedure will be used here as the recommended standard. When working to this, designers should take account of the other, related codes that complete the Co-ordinated Project Information initiative.

The production drawings code begins:

'The code is about the drawings that appear on every building site and contain drawn information for construction. Throughout the code they are referred to as the production drawings or collectively as the production set and include those drawings produced post tender by specialists and sub-contractors.

An effective set of production drawings cannot be produced simply by following a single, tried and tested routine. Guidance must consist of firm advice, some guidelines and discussion of options.

For every new project, decisions have to be taken about such matters as co-ordination, arrangement, format and content of the production set. Many of the decisions can only be made in the light of the particular circumstances and the code is designed to assist in making those decisions and to show how these fit into a programme for preparing and issuing the drawing set.'

The code includes the following, some of which are expanded below.

- Co-ordination of information on drawings.
- Drawings arrangement.
- Format of the drawings set.
- Example drawings arrangement.
- Drawings content.
- Planning, preparation and issue of the drawings set.
- Drawn information at tender stage.

8.1 Co-ordination of information and approval of inputs

The current situation over responsibility and liability for design makes the task of producing a set of co-ordinated information complex. Despite the basic assumption that all design is the architect's responsibility, the complexity of modern construction technology has led to an increased dependence on specialists at both the design and construction stages. As a result, the client will often give authority to an architect to delegate specified parts of the design to others, and it may even be implied from the circumstances of the case. In recommending the appointment of a particular specialist, architects owe their clients the normal duty to use reasonable care and professional skill. Although they may not automatically be directly responsible for the defaults of the people they recommend, their overall obligation to co-ordinate the design will render them at least partially liable for all design defects.

In the UK, the Construction (Design and Management) Regulations make it incumbent on clients to employ only those who are reasonably competent. Moreover, this is a criminal, not civil liability. This obligation also applies to architects who sublet parts of the design to specialists. In terms of producing information, these regulations oblige the client to appoint a Planning Supervisor, to notify the project to the Health and Safety Executive and to compile a Health and Safety File, in parallel with the design process. This file must be handed to the contractor when construction work commences, and ultimately to the client.

Apart from recommending specialists, an architect's duty of care and skill involves the responsibility for directing and co-ordinating the expert's work into the project as a whole. Thus, if any danger or problem arises in connection with the work allotted to an expert, of which a competent architect ought to be aware, then it is the duty of the architect to warn the client.

The project manager and design managers should therefore decide, at the outset, who has the primary responsibility for ensuring that the content of the drawings is co-ordinated. It can be the architect, the lead designer, or a mutual responsibility placed on everyone. Whichever it is to be, it should be clearly stated in the agreements of all the parties including those of the contractor and the works contractors. Two aspects of co-ordination need to be differentiated.

- The need to produce information that fully interconnects the inputs of all the contributors into one coherent and complete piece without ambiguity.
- Ensuring that all contributors are working to a co-ordinated schedule to achieve a timing of the information flow that allows the development of the co-ordinated information.

It may be necessary to produce specific co-ordination drawings that put interrelated information onto a single drawing. Techniques such as overlay draughting and CAD are suitable for this, but only the most sophisticated CAD systems include clash control, which automatically identifies situations where, for example, pipes pass through beams. However, this is typical of the situation that must be identified before the work starts on site. All too often the contractor is faced with the problem of collecting information from a variety of drawings produced by the design team and works contractors to determine how an assembly is to work. Problems at this stage lead to emergency decisions that may produce expensive and less than optimal solutions.

A final issue, perhaps the most important, is that of the provision of accurate information on the position and size of holes. In all structures the forming of holes is far cheaper than cutting holes. Therefore, it is essential that all services locations and penetrations of the structure are identified. Two techniques are commonly used.

- A policy of zoning for the location of holes is established prior to the structure being fabricated or constructed. The zones can either be left as large holes or of 'soft structure' that can easily be removed when the precise locations are identified. This approach allows the work to progress before all specialist design is complete.
- A set of layout plans can be circulated to all designers requiring them to indicate the location and size of holes. This must be checked for structural integrity and effect on the design concepts before issue to the site. All design inputs, including specialist trade design, must be available for this method to work.

8.2 Review of design input

Given that the specialist contractors will be contributing design information, the role of the architect in incorporating their information can be ambiguous. The recommended terms of engagement, given by the RIBA, require the architect to review the design of others. This is adequate where the specialist's design is self-contained and isolated.

However, where the specialist is providing components, which are incorporated into other assemblies, and where possible failure is not within the component but its application, then the architect must take responsibility for assembly design.

A common practice, particularly in the USA, is to use an A, B or C action category of review, where a specialist's drawing is classified as follows.

- *A action* – The contractor may proceed to authorize fabrication, manufacture and/or construction, providing the work is in compliance with the contract.
- *B action* – The contractor may proceed to authorize fabrication, manufacture and/or construction, providing the work is in compliance with the designer's comments and the contract.
- *C action* – The contractor shall not proceed to authorize work to be fabricated, manufactured and/or constructed. The contractor shall re-submit revised documents.

This system can be effective if the specialist is given help to understand why there are queries on the drawings. However, in practice there are many examples of the majority of the drawings being classified B, thus placing an ambiguous risk on the specialist. The project and design managers should keep a close watch on this type of approval system to ensure the continuing co-operation of the specialist.

8.3 Computer-aided design

The computer is a powerful tool that is used increasingly as a drafting and co-ordination device. Productivity in terms of drawing production can be multiplied by a factor of two, given experienced users. One major advantage for designers is the ability to change and modify the model of the building rapidly and so test alternatives.

The advances in co-ordination are very dependent on the mechanisms of data exchange between the contributors to the design. Where the design team is multi-disciplinary and has access to a common system, there will be considerable advantages, although as already mentioned there are, at present, few clash control programs available. For CAD to be used effectively, the designers and project managers must determine a strategy that maximizes the advantages.

In many cases the individual design organizations will have invested in their own CAD systems and will be familiar with their operation. However, data transfer between them will have to be undertaken by paper in the normal way. This has the inevitable drawback of additional effort spent in transcription with the dangers of error creeping in. Electronic data transfer between different systems is an ideal situation. On larger projects it may be worth considering investing in a common system for all the primary design participants either through a CAD bureau or by using the system of one of the participants as the primary CAD system.

Considerable research is under way to take advantage of the data transfer protocols, such as STEP, to achieve interoperability between open systems. The first application of this in the construction industry has been CIMsteel (Computer Integrated Manufacture of Steelwork: http/www.leeds.ac.uk/civil/research/cae/cae.htm). CIMsteel has demonstrated the integration of the complete design process from outline design through to manufacturing details via the seamless transfer of data from one computer system to another. There is no loss of data from one system to another and the results of calculations in one system can be used by another system and vice versa. The potential is to go from outline design to details in less than 2 hours with the confidence that all of the data is compatible and there is no data corruption when transferring from one system to another. All of the main CAD vendors are now developing translators using STEP and once this is generally available there will be little need to invest in customized project-based IT solutions. This area is advancing rapidly and the introduction of transfer via the world wide web with the introduction of XTML (Extended Text Mark-up Language) may supersede the work of STEP.

However, even with the implementation of interoperable systems there is still the need to communicate between remote users to discuss issues that involve both dialogue and the manipulation of the CAD data and drawings. The introduction of world wide web technology and wide band communication is allowing the use of teleconferencing and also interactive video CAD. Designers can view each other and also the design information on the same screen and actively discuss changes that each is making to the information in the CAD model. This is highly attractive in that it removes the need for co-location, but as yet there are no large experiments of total project teams integrating in this way. Whether there will ever be the possibility of producing a design without the need for face-to-face dialogue is an interesting question, which time and the development of technology will answer.

A major practical hurdle to the use of fully integrated design software is the decision as to when the information is sufficiently complete to issue for construction. A common complaint is that, whereas the status of the design produced by a draughtsman could be seen, when it is in a computer system with limited access it is far harder to see the status. An intermediate measure is to issue paper copies at regular intervals so that the status can be judged by traditional means.

8.4 Issuing information

To assist in the supply of complete information, the drawings should be issued in sets bound for either the zone, the system in a zone, or the component within the system in the zone. When it is issued the lead designer for the work should authorize the set as complete and fully co-ordinated. This simple discipline is intended to avoid error.

Another problem to avoid is the separation of information into different media. For example, information on drawings is often issued separately from the relevant specification. However, it is common practice in the USA for specification information to be produced on self-adhesive, transparent sheets that are added to the drawing film. The reader of the drawing then has all the information needed in one place for quick and accurate reference. This also saves time in the drawing office, as the specification can be produced faster by word-processor than by hand. A CAD system with a database facility would achieve the same solution.

It is generally thought that project size determines the way in which drawings are referenced and this approach has been adopted in the CPI. However, only a referencing system orientated by location would embrace the zone-related philosophy of organization advocated in this book (Table 8, see page 109). Therefore, if the zone-based organization of the design process is followed, the drawing referencing should adopt the same pattern.

8.5 Drawing control

Each organization must keep a precise record of the information it generates, its current status, the distribution and, where appropriate, the stage it has reached in the approval system. For simple projects and small offices, a paper-based record system will suffice, however, on large projects a computer database is desirable.

8.5.1 Computer record systems

There is a need for a record system that covers all drawings from all contributors. An initial decision should be made about which organization has responsibility for maintaining the project record. As it will also be used for progress chasing, it should be the responsibility of the project manager or the design management.

Most record systems use a computer database developed to the specific needs of the user organization and not necessarily the needs of the project as a whole. The features required (see Figure 26) enable the development of a historical record for each drawing, showing its development, approval and timing of distribution through all stages. The more sophisticated programs will link into planning systems for a target date against which progress can be measured. This allows exception reporting for progress control and notes of the course of action chosen. By breaking down the reports into disciplines and zones of work, a precise trail of responsibility for progress can be kept.

8.5.2 Drawing storage and reprographics

Most projects generate a vast volume of drawings and other information that is difficult to organize, file and keep in good condition. Electronic archiving (based on scanning

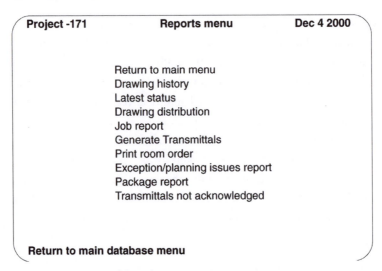

Fig. 26. Typical screen menu for integrated drawing management systems.

source documents) is becoming commonplace. If this is considered at the outset, it can be linked to the original distribution system. If all drawings are CAD or scanned before issue, then the printing system could be operated by a reprographics department or agency by just calling up the record and issuing the distribution instructions. If all demands for drawings are dealt with in this way, a central control system is created that should be more efficient. Moreover, this will reduce the scope for misunderstandings about the issue and receipt of drawings. If this idea were extended to a project-based system for all drawings issued, including trade contractor drawings, it could be economic to have a central reprographics system for the project.

8.5.3 IT systems for drawing control

The next logical step from microfiche and central reprographics is to use computer-based data storage. This system allows differing media for drawing creation to be used which are then combined into one set of digitally held information accessed in a variety of ways. At the heart of the system is a high capacity storage medium, on to which all of the project information is scanned with a reference code. Drawings can be accessed from remote terminals, edited and commented on for reading by the issuer, issued for use and even printed locally as needed.

This type of networked, uniform, digital, drawings database and centralized document control gives the capability to access, distribute, comment and archive all on one system. The key benefit of the system is the distribution of the information at the speed required to keep the project moving. The speed of information retrieval is increased, thus eliminating the time spent searching for drawings, letters or instructions. The drawing control database logs details of all drawings issued and automatically allocates the relevant distribution using the distribution matrix number assigned to each drawing. Transmittal letters and acknowledgement slips are produced for each recipient. The issue can be tracked until receipt of signed acknowledgement slips or drawings, which were issued for approval or comment, have been returned.

This type of system is a considerable step forward until integrated CAD systems provide for full data interchange between all contributors to the design as well as to the site.

9

Evaluation of information

The independent review of the content of the design (to ensure that it is competent, gives value for money, and will provide long-term satisfaction in use) is a requirement of modern design and must be approached in an orderly way. This chapter includes descriptions of some of the reviews that should be undertaken formally during each stage of design.

9.1 Design reviews

The convention of many design teams is to hold an internal 'crit' session of the design at key stages, such as initial concept and scheme design. These are critical reviews involving people both working on the project and from other projects. The object is to view the developing design, to share the ideas evolving within it, and to receive alternative views in order to improve the design.

A more formal system is that required by ISO 9004 (British Standards Institution 1987) which requires that design reviews be carried out at the conclusion of each phase of design development by conducting 'a formal, documented, systematic and critical review' of the design results (Cornick, 1990). The reviews should 'identify and anticipate problem areas and inadequacies and initiate corrective actions to ensure that the final design and supporting data meet client requirements'. The review process is divided into categories, which, depending upon the stage, can be interpreted as appropriate.

Items pertaining to customer needs and satisfaction include the following.

- Comparison of client needs with standard or innovative technical specifications for materials, products or processes.
- Validation of the design by building product or element prototype tests.
- The capacity of the designed building to perform under expected conditions of use and environment.
- Consideration of unintended uses or misuses by potential users.

- Building safety and environmental compatibility.
- Compliance with building regulations, planning regulations, health and safety law, national and international standards, and designer's corporate practices.
- Comparisons with other practices' comparable building designs.
- Comparisons with own similar designs, especially analysis of own and other practices' external problem history to avoid repeating past problems.

The following items pertain to product specification and in-service requirements (these relate to the engineering and specifying stages and, if carried out in full, should assist in the prevention of errors).

- Reliability, serviceability and maintainability of the building structure, fabric and services components.
- Permissible assembly tolerances and comparison with process (material/methods of assembly) capabilities.
- Buildability; ease of assembly, storage needs, component life and disposability.
- Aesthetic specifications and acceptance criteria.
- Failure modes and effects, and fault analysis.
- Ability to diagnose and correct problems.
- Labelling, warnings, identification, and traceability requirements of building components or elements.

Items pertaining to process specification and in-service requirements include the following.

- Buildability of the detail building design, including special process needs, mechanization and automation of the installation of building components.
- Capability to inspect and test including special test requirements.
- Specification of building materials and components, including approved supplies and suppliers.
- The safe execution of building operations, especially operative safety and working conditions.

9.2 Value engineering

All designs that are the result of extensive parallel working can benefit from a 'second look'. Value engineering (as opposed to Value Management, see page 98) is a systematic approach that seeks to enhance value by eliminating unnecessary cost while maintaining function (Green and Popper, 1990). The greatest return from a value engineering exercise is during the early design stage and before 60% of the design commitment is made. When value engineering occurs later it may be treated as a cost cutting exercise and will probably be undertaken in a hostile climate that will prevent it from achieving the desired objectives.

A method often adopted is to hold a 40-hour workshop, in a location separate from the normal workplace, where the key members of the design team can focus upon the significant issues of value, usually with the help of a consultant specialist in value engineering. Another approach is to assemble an alternative design team that would undertake the critical review of the design. The merit of the existing design team undertaking the review is that it increases their involvement and commitment to the project through the intensive working sessions with all the key decision-makers.

9.2.1 Value engineering methodology

The recommended methodology has five stages.

- *Stage I: Information* – The start of the study is the isolation of the most significant items in terms of cost, programme or quality, and the identification of their function. Typical prompt questions used to isolate function include:
  - what is it?
  - what does it do?
  - what else does it do?
  - what does it cost?
  - what is its value?
- *Stage II: Speculation* – This phase is used to generate alternative ideas. The key to its success is the freedom to think widely, creatively and not to be inhibited by convention or previous knowledge of the project.
- *Stage III: Evaluation* – Once a reasonable range of alternatives is proposed they need to be evaluated against the criteria for value and their suitability for the particular project.
- *Stage IV: Development* – The most significant alternatives are then developed in detail. Any solutions that do not provide increased value or reduced cost either in the short-term or over the life of the project will be rejected.
- *Stage V: Presentation* – Where an alternative design team has been used, they will have to present their ideas to the existing team so that they understand the thinking and significance of the alternative approach.

Value engineering is a positive contribution to the design process, but its timing is crucial to avoid delay to the project or large amounts of abortive work.

9.3 Buildability

The site production activity is dependent on the detail of the design. The increasing use of design by works contractors is one way of obtaining simple, easy to build details because contractors can apply their production experience to the details that they propose.

As long as the bid documents and the process of detail design are open enough to allow for adjustment of the details, then a buildable design will emerge. However, if the detailing cannot be left to a works contractor to make a full trade-off between production, cost, time and quality, then it will be necessary for the design team as a whole to consider buildability as an integral part of their activity.

While the design of simple-to-construct details and connections should be the task of all designers, the complexity of construction operations is difficult to determine without extensive site experience. The buildability checklist (see Checklist 12) gives a guide to the steps in the evaluation process and the strategic issues to be considered in each step. Most benefit is gained if the buildability evaluation is undertaken at the earliest opportunity, perhaps as a contribution to the concept stage.

Experience has shown that the essence of the buildability decisions, taken in any earlier strategic analyses, must be maintained throughout the detail design. For example, pre-cast concrete cladding panels can be an economic form of cladding, but their cost will rise dramatically if a large number of different panels are needed. Even small variations in basic dimensions, the insertion of windows, or the addition of decoration will rapidly increase the number of 'specials'. Much of this kind of detailing is added during the detail design process, and can rapidly destroy the economies of prefabrication implicit within the initial strategic decision.

9.4 Hazardous operations

In operations that involve complex processes, dangerous substances, and potential safety risks, there is a need for a particular review of the design by those who are to operate the building and its processes. This needs to be done in increasing detail as the design progresses. However, it must not be left too late as major changes during the construction stage will lead to delays and major cost increases. In the process industries, considerable effort is put into this review, with scale models of the plant used to test both process and escape procedures. Checklist 13 is a guide to the main areas of review.

9.5 Life cycle cost

It was not until the early 1960s that it became common practice to set up formal procedures to plan and control the capital cost of construction work as an integral part of the design process. The methods that were employed enabled informed choices to be made regarding the cost of alternative design solutions (at both outline and detailed levels) and their effect on the cost of the development as a whole. Within a comparatively short time, the success of

Checklist 12.   Buildability checklist

1 Calculate a construction programme:
- determine a sequence of activities
- identify key interfaces
- recognize patterns of usage of significant or key resources
- determine realistic time periods given complexity of operations.

2 Evaluate the construction time:
- assess the time in relation to other similar projects
- produce more detailed analysis of long or complex sections
- identify intermittent patterns of resource use
- identify activities that control the speed of construction.

3 Re-evaluate design to avoid:
- bottleneck activities
- slow speed activities
- inefficient use of resources within activities
- inefficient use of resources over several activities.

4 Re-design to achieve:
- continuity of resources at a constant and minimal level
- utilization of recognized and traditional skills
- the use of the commonly available level of skill
- the maximization of construction plant and equipment
- sensible prefabrication of complete sections of the work, e.g. toilet pods, lift motor rooms, plant rooms, service risers, and shaft walls
- practical tolerances between components.

these procedures has led to attempts to bring the longer term running and replacement costs into the calculations. By the end of the 1980s the term 'costs in use' was current, but was soon superseded by the more self-explanatory title 'life cycle costing'. Other terms that have been coined, some of which have fallen by the wayside, include 'terotechnology', 'whole life costing', 'through life costing', and 'total costs of ownership'.

The basis of life cycle costing involves putting the estimated capital, maintenance, operating and replacement costs into a comparable form, and bringing them together into a single figure which allows for the fact that these items of expenditure will take place at different stages within the time-scale (see Checklist 14). There are many techniques for doing this.

Checklist 13.   Hazardous operations checklist

Hazardous processes and materials:

- storage
- structural stability and resistance to the leakage of the material
- resistance of the finishes to the material
- spillage control – bunds, temporary storage, access for clearing equipment, air quality control, emergency ventilation
- materials handling
- equipment – machinery, power, servicing, maintenance, control rooms
- processing
- control systems – fail-safe operation, back-up systems, system redundancy, spillage control, air cleaning, safe residue control and disposal
- packaging/storage/shipping.

Operative safety:

- emergency evacuation
- refuges, routes (air locks, lighting, air supplies, walkway widths, rescue services access)
- protective clothing
- storage, changing, cleaning, disposal
- breathing equipment
- storage, compressors, maintenance
- maintenance procedures
- communications
- rescue and control team facilities.

Building safety:

- containment zones
- safe air, vapour and liquid containment and evacuation under control to protect the environment
- pressurized air zones
- air handling systems linkages
- door controls/air flows
- alarm systems
- fire fighting systems
- active and passive.

Checklist 14.   Checklist for life cycle cost analysis (from Flanagan and Norman, 1983)

For each element:

- capital cost
- anticipated life
- running costs
- operations costs
- annual maintenance costs
- maintenance/replacement/alterations (intermittent)
- tax allowances
- inflation rate
- inflation adjusted discount factor
- obsolescence factors:
    - physical
    - economic
    - functional
    - technological
    - social
    - legal

Strictly speaking, life cycle costing is concerned only with the costs of an asset. However, if there are any benefits that can be expressed in monetary terms then they should also be taken into account, as otherwise only a partial picture of the cost-effectiveness of the asset will emerge. If this is done, then the exercise is more properly termed investment appraisal, although it is considered as part of the field of life cycle costing.

These techniques enable straightforward comparisons to be made so that the most cost-effective approach can be chosen as the solution to the problem. Almost any form of a tangible asset, ranging from a simple floor covering to a fleet of nuclear submarines is, in theory, amenable to this treatment.

Life cycle costing has gained favour as an idea. It is inherently more sensible to look at the total cost consequences, rather than just capital costs, when looking at the choice between alternatives.

9.6 Maintenance

The design should allow the maintenance of the building to be carried out in a safe and reasonable way. In the UK, this is a legal requirement, imposing obligations on clients, designers and builders. The design review is undertaken to ensure that all parts of the building that will need to be maintained during its life can be accessed safely and with the least disruption to the users.

The first issue is to examine each component and assembly, determine its design life, and then examine the details to ensure that the design life will not be reduced through poor attention to detail. This analysis will make many assumptions about the way that the building will be maintained. These should be recorded to form the basis of the maintenance manual for the building and they need to be incorporated into the Health and Safety File.

There is another legal issue concerning safe access for maintenance. The Health and Safety at Work Act 1974 places an onus on the employer for the safety of maintenance workers, i.e., the owner or occupier of the building is responsible for the safety of any employee in the building. This includes visiting maintenance workers (Stavely, 1991). There are three main issues to be considered: access for inspection, external maintenance, and services maintenance.

9.6.1 Access for inspection

Regular inspection of the key parts of the building by the maintenance team will be required, and inspection covers, doors into interior spaces, access behind equipment and on to roofs must be provided to ensure that there is no barrier to the inspection.

9.7 External maintenance

Powered cradles are a common feature of many large buildings and are essential for regular cleaning and inspection. Smaller buildings without this provision will need scaffold access that may need tying into the building. Permanent ties should be provided. Access to roofs should be from inside the building where possible. For non-load-bearing walls, access to the fixings will be required for regular inspection to ensure that they are not corroding.

9.8 Services maintenance

There are three main areas of services maintenance to consider.

- All services must be zoned to enable areas to be taken out of service for routine maintenance without closing down the whole system.
- Access to services must minimize disruption to users and also not cause any damage or deterioration to finishes. For example, access to services above suspended ceilings for

adjustment of controls should be through access panels and not through individual tiles.

● Equipment replacement should be considered and doors should be provided which are wide enough to enable the equipment to be removed and new equipment delivered.

The design team must consider how the building is to be maintained and must compile the maintenance manual as the design progresses. Handing over the building to the client should involve educating the operating and maintenance organization that is going to manage the building. Sufficient time and money should be built into the budget for this activity. Even where the client is going to subcontract this activity, considerable training of the operators will be required. The Construction (Design and Management) Regulations lay down detailed obligations about how this information should be compiled and the timing of its transfer to those who are to use it.

10

Planning, monitoring and control

The planning of the design activity is fundamental to design management. A different approach must be considered for each stage of the design. At the outset there is a need for a strategic overall plan which considers all stages of the work, the interface to the construction process, and the activity of the key contributors to the design, including the works, specialist and trade contractors.

As the scope of the project is gradually determined, the strategic plan can be refined as the scale of the contribution from each of the members of the design team and the specialists becomes clear. Once the agreed scheme design stage is reached, the planning becomes very detailed and should include every significant interaction between all designers, including every works and specialist contractor. This must be done, stage-by-stage and zone-by-zone, by the people responsible for the work and co-ordinated by the lead designer for the zone.

At the finest level is the requirement for a schedule of all drawings to be produced by each designer, which must reveal the linkages to other, interrelated, design work.

10.1 Methods of programming

The following methods of programming are commonly used:

- network analysis
- bar charts
- procurement schedule
- information-required schedule
- information-transfer schedule.

The key to their successful use is whether the underlying framework of crucial activities, which provide the essential planning and control information, can be easily presented and interpreted.

10.1.1 Network analysis

Design is a complex network of the interchange of information, but experience has shown that a critical path network, while useful for project control, is not a suitable form of programme for everyday use by the majority of designers. The strength of the method is its ability to show the essence of the interactions between the work of each designer. However, as the work increases in complexity, particularly in the engineering stage, the volume of the interactions and their complexity can easily swamp a network, even for a small project. Network analysis, particularly computer processed, requires great precision in reporting and data entry. This leads to great difficulty in providing the flexibility needed to cope with, for example, design iterations and partial completion of information.

10.1.2 Bar charts

Bar charts are the most common form of programme because they are very simple. They are most effective in the early strategic phases of the work to show the approximate disposition of the main sections. They clearly illustrate the activities to be done, but have limitations in showing complex interconnections between activities. Also they are not very effective for showing the plan for the production of large numbers of drawings and details, as it is difficult to keep them up-to-date due to the complexities of information interchange and iteration, particularly on a project wide basis.

Individual bar charts of up to 30 activities are quite effective, but with this approach there needs to be an overall co-ordination of the various programmes to ensure consistency and the awareness of the necessary interactions between the individual sets of activity. At the more detailed level of information production control in an individual office, or on a designer or zone basis, the bar chart is effective (see Figure 27). The 'percentage complete' scale is only a guide, but those who use this method find that it works reasonably well for identifying problems quite quickly. The interconnection of design information between domains is difficult to portray and therefore percentage completion of a particular drawing has little relevance as an absolute or accurate measure on overall project performance.

10.1.3 The procurement schedule

This is a schedule often seen on management types of contract, where each package of work and the stages to achieve its procurement have been identified. It is an extremely useful document as long as certain conditions are met. Lead times to the construction are determined by two things: first, the manufacturing period for the components, which in some cases can be long; and second, the specialist's shop drawing time and the need to obtain information from other designers, some of which may be other specialists. If this

Drawing production programme

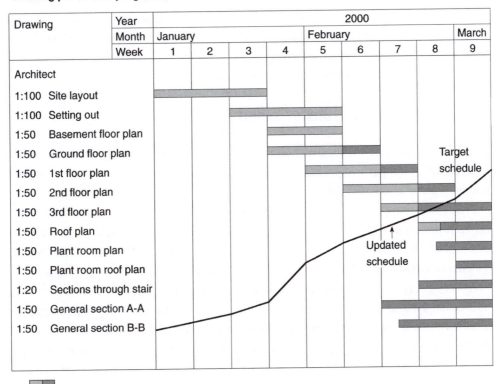

Fig. 27. Sample bar chart-based progress monitoring of drawing production.

is the case then the lead time will be determined by the interchanges during the shop drawing stage.

10.1.4 Information-required schedule

The most common design management tool is the 'information-required schedule' which is prepared by the contractor at the start of the job to assist the design team in understanding the production information requirements. The drawings and information are listed together with a date for their receipt. This is monitored, usually monthly, at the site meeting or design progress review meeting.

This schedule is often issued too late to influence significantly the production of complex sequences of information, particularly if design input from nominated or domestic subcontractors is required, so delay may well occur. The design team should have its own production plan drawn up months before contracts are placed, which is sympathetic to the needs of the construction process.

10.1.5 Information-transfer schedule

A method of planning the design process has been developed to overcome many of the deficiencies of the other methods. This recognizes that there is a network of transfers of information and at the same time does not attempt to monitor the production of each drawing. It divides the information into sets of documents and information that are transferred between participants (see Figure 28). The sets comprise drawings, sketches, specifications and text. It is the role of the recipient of the information to establish the quality of the content, which is obviously related to the task to be performed once the information is available.

As design is not easily produced to a strict time-scale, because of the variable intellectual input and the need to develop ideas over several iterations, the programme must allow the designers freedom to use their initiative, while maintaining an overview of the state of progress. The programme is a vertical bar chart for each design discipline, split into zones and systems. Each discipline has a sheet so that all the information that they need is readily available without cross-referencing. The planner or manager is the one that ensures that each sheet picks up all the input and output from each designer. The crucial aspect is that the designers should be made aware of the importance of their output in terms of the way that it forms input to the work of others. This is shown in the expanded text to interpret the programme requirements in the lower half of the programme sheet. The architect, in this case, was required to supply information on 10/3/99, which, as it was done to the satisfaction of the structural engineer, has been shaded in as complete. The text describes the importance of the information to the structural engineer, thus giving the architect a clear reason for the interchange. The recipient specifies and becomes the arbiter of the quality of the work and is the person who, at the review, states whether the programme has been met. This method gives the designers:

- the freedom to determine a work pattern related to the importance of the particular piece of information;
- an ability to adjust their own priorities within the time-scale;
- a degree of flexibility that is compatible with the designers' inherent freedom.

At the same time, the method ensures that the flow of information is carefully controlled to ensure that each designer is not delayed through the failure of others to appreciate the information needs. This method can be extended to all who make decisions, such as clients, contractors and external bodies.

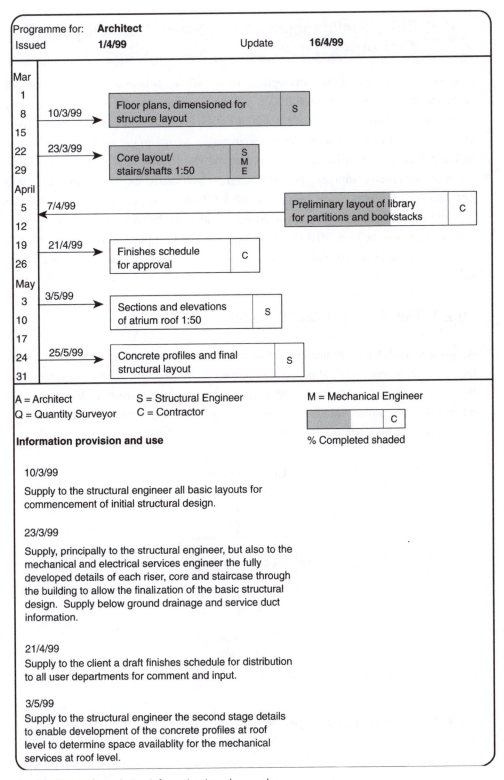

Programme for: **Architect**

Issued **1/4/99** Update **16/4/99**

Mar	
1	
8	10/3/99 → Floor plans, dimensioned for structure layout [S]
15	
22	23/3/99 → Core layout/ stairs/shafts 1:50 [S M E]
29	
April	
5	7/4/99 ← Preliminary layout of library for partitions and bookstacks [C]
12	
19	21/4/99 → Finishes schedule for approval [C]
26	
May	
3	3/5/99 → Sections and elevations of atrium roof 1:50 [S]
10	
17	
24	25/5/99 → Concrete profiles and final structural layout [S]
31	

A = Architect   S = Structural Engineer   M = Mechanical Engineer

Q = Quantity Surveyor   C = Contractor

Information provision and use   [C]

% Completed shaded

10/3/99

Supply to the structural engineer all basic layouts for commencement of initial structural design.

23/3/99

Supply, principally to the structural engineer, but also to the mechanical and electrical services engineer the fully developed details of each riser, core and staircase through the building to allow the finalization of the basic structural design. Supply below ground drainage and service duct information.

21/4/99

Supply to the client a draft finishes schedule for distribution to all user departments for comment and input.

3/5/99

Supply to the structural engineer the second stage details to enable development of the concrete profiles at roof level to determine space availablity for the mechanical services at roof level.

Fig. 28. Sample for a design information interchange plan.

10.2 The development of an analytical design planning tool (ADePT)

Research is continuing in the development of specific tools to assist in the understanding of the interdependencies between design activities and to deal with the iteration necessary to produce co-ordinated information. A multi-stage approach has been developed (Austin *et al.*, 1999a). It is aimed at the detail design stages where, in the analysis of four typical building designs, the building design process is shown to comprise between 7 and 10 iterative loops each comprising between 5 and 30 interrelated loops (Austin *et al.*, 1999b). The number of design tasks averages around 350–400, but the number of information dependencies is over 2400. On larger projects, with over 750 design activities, the design dependencies can be over 10 000. Clearly, not only is it important to identify the design activities, but also the information interdependencies. The ADePT approach takes four stages.

10.2.1 The design process model

A process model that represents information requirements, i.e., constraints, without indicating how each task should be undertaken, is used. By using a method that is designed to separate the inputs from three sources (inputs from within a discipline, interdisciplinary inputs and the inputs from external sources), a more realistic model of the constraints can be made (see Figure 29) (Austin *et al.*, 1999c).

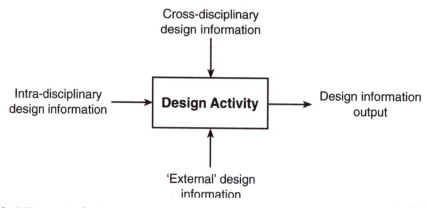

Fig. 29. IDEF$_0$ notation for design process maps.

10.2.2 An information dependency table

The model generates a list of activities, associated with which are the information constraints. Each constraint will have its own sensitivity to the stage and development of the knowledge of the situation. Thus, at each review, the information can be classified depending on its relative importance to the current work.

10.2.3 Dependency structure matrix analysis

A matrix of the design activities is created and for each activity the dependency is shown together with the relative strength (see Figure 30). The activities are initially arbitrarily listed in sequence down the left and the dependencies shown against the same list across the top. The matrix is then re-scheduled to ensure that the maximum number of activities do not have dependencies to the right of the diagonal. Where this is not possible then iterations will occur in the process until the information matches the requirements.

10.2.4 Project and discipline design programmes

From the matrix manipulation the changes in the process ordering are identified and a design plan can be produced where the time-scale for each stage can be developed and a programme produced.

The advantage of this approach is that there is a very clear emphasis on ordering the design stages to meet the requirements on information interchange. Consequently there is an attempt to minimize the iterations commonly seen in design processes where the interplay between members and their knowledge has not been fully considered and structured. The tools from this research, once developed, will reveal the structure beneath the information transfer.

10.3 Work package design and procurement control

It is quite usual for the sequence of construction activity on the site to be used to generate the programme for the information production, whereas the information interchange requirements demand a different sequence to be established. In component construction, there is considerable interchange between the designers and specialist contractors to resolve details at the junctions between components. The need to bring this information into the process often requires the specialist contractors to be appointed in a different sequence from their normal pattern of site work to ensure that their special knowledge is brought into the design process at the right time. The interchange between specialist contractors must also be appreciated and should be included in the network of design activity.

To ensure that the design is progressing fast enough to allow the placing of the work packages with the contractors at the right time, a simple works package progress

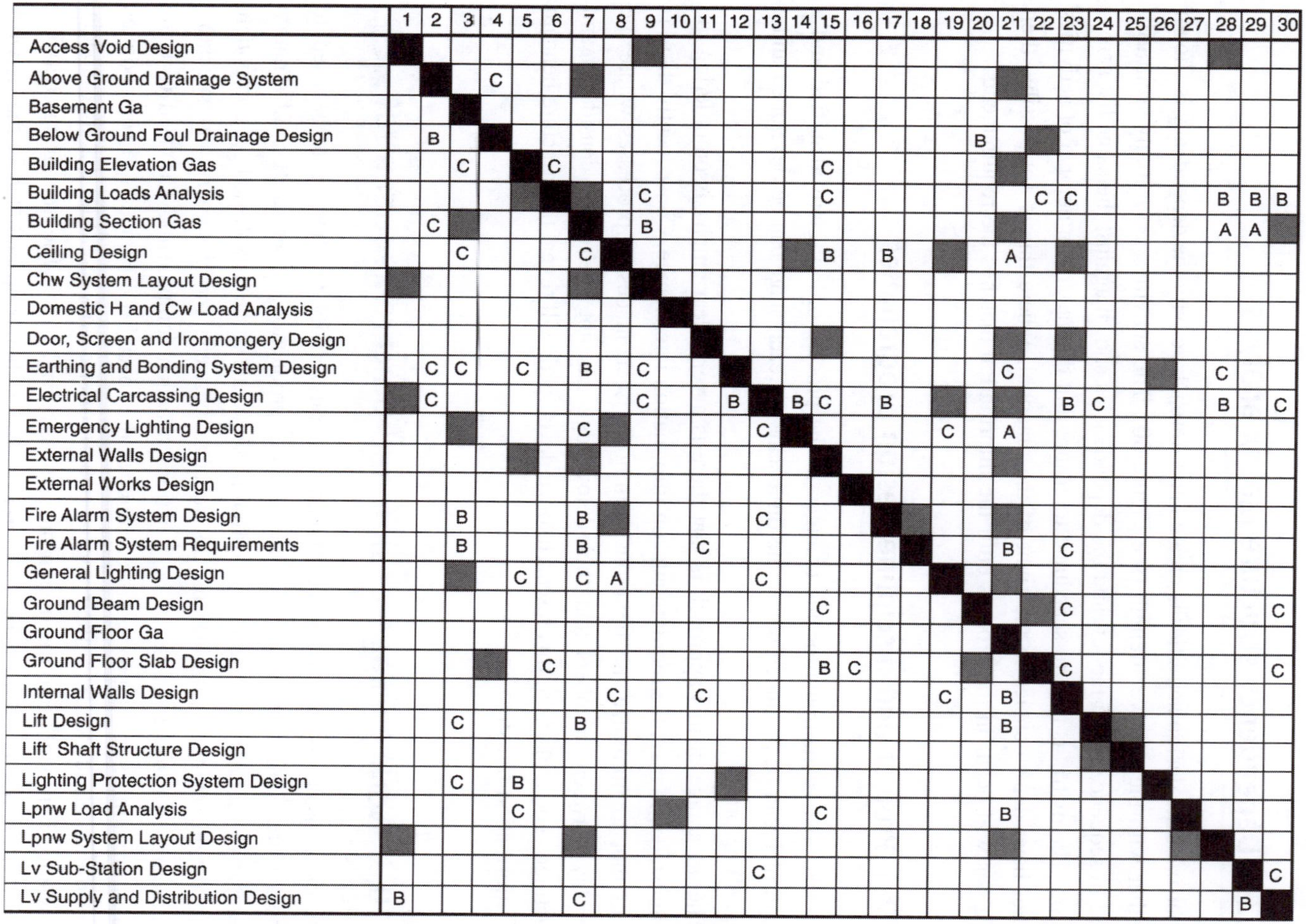

Fig. 30. Prioritization and missing information identification from an ADePT matrix.

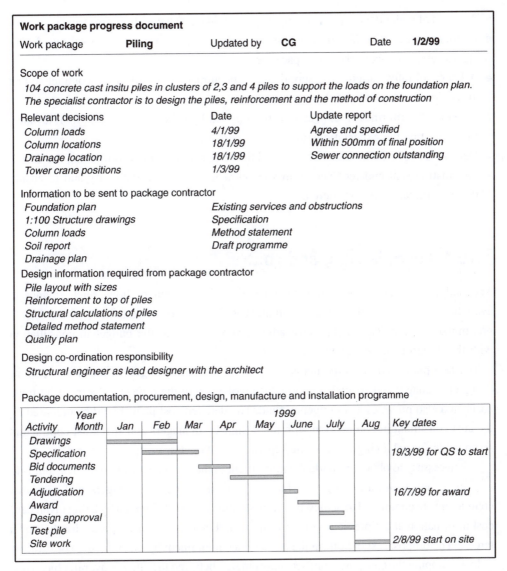

Fig. 31. Sample for a work package progress document.

document can be used (see Figure 31). The objective of this document is to keep a record of all decisions relevant to the work package, as well as a record of planned progress. The primary task is to avoid ambiguity and aid communication of the decisions that affect the work package.

The work package progress document shows the following:

- The scope of the work by a concise description of its key features and any departures from normal expectations.
- A programme of the major activities leading to the placing of the work package contract.

- A record of major decisions affecting the programme that have arisen from any revisions.
- A brief record of all design decisions made during the progress of the various design stages, which concern the work package.
- A list of all information to be supplied to subcontractors as part of the contract so that all designers are aware of their contribution.
- A list of the information expected to be supplied by subcontractors and its relevance, to ensure that the expectations of the designers are met.
- Decisions about the interface control and the responsibility for co-ordinating the design information sent and received from subcontractors to avoid ambiguity and to ensure that co-ordination is considered.

10.4 Cost planning and control

The budget may either be based on a developer's decision of how much finance is available, or alternatively, on an estimate based either on a statement of need or a preliminary design. The client's cost adviser should prepare the budget in consultation with the design team leader and the client.

The cost plan subdivides the budget to provide a separate target for the work of every designer, manager and specialist contractor involved with the project, on a trade-by-trade basis, and also provides a contingency fund. The detailed cost plan should be prepared by the cost consultant in conjunction with the designers, managers and specialist contractors responsible for achieving each of the separate cost targets.

It is necessary to obtain individual commitment to the separate cost targets. The aim is to set tough, but achievable, targets, which individual designers, managers and specialist contractors are committed to achieving. They should each be challenged by being given hard cost information about previous similar projects and then asked for proposals for their own targets. The proposals should form the basis for a commitment by the individual or firm.

In developing the cost plan, conflicts may emerge between the client's overall objectives, design considerations and the dictates of good management practice. Where any such conflicts arise, the client must decide how the budget is to be allocated in the cost plan.

The cost plan should be a living, evolving document that is kept up-to-date with the current set of decisions about costs. It should be available to all project participants as a usable control document for the whole project. The cost consultant is responsible for keeping the cost plan up-to-date and for ensuring that it communicates the importance of cost control to everyone involved. It is important that everyone involved with the project feels responsible for cost efficiency. This is possible only if they understand the budget and cost plan. The cost adviser or project manager should arrange seminars on the project cost planning and control process to ensure that everyone understands the figures that apply to them.

It is important that cost targets are co-ordinated. Obvious examples of co-ordinated decisions include the following.

- Ensuring that all cost figures make consistent assumptions about levels of cost and inflation.
- Ensuring that assumptions about the boundaries of work items and the provision of plant, equipment, storage, distribution, welfare, etc. are consistent.
- Ensuring that the cost of specialist contractor's design is included.
- Ensuring that assumptions about the management of the project, such as responsibility for unloading, distribution of materials, clearing away rubbish, etc., are consistent and fully allowed for in individual work items.
- Ensuring that the cost of temporary works and the method of operations have been fully priced against each work item.
- Ensuring that consultants' fees, land cost, and finance costs are fully understood and considered.
- Ensuring that the contingency allowance in the cost plan is managed consistently.

Everyone is responsible for co-ordinating his or her own decisions with the rest of the project team and so should be made responsible for discussing all assumptions with everyone else likely to be affected. Everyone should be active in containing costs within his or her own cost targets and in looking for ways of improving efficiency without reducing quality.

It is necessary to maintain an up-to-date record of costs committed against each cost target, so it is important to have fast feedback on changes. It is much better to have an approximate report on costs today, than to have an exact figure too late to take any necessary corrective action. Each individual who is responsible for a cost target should keep his or her own cost commitment record. These should be collected, reconciled and turned into a monthly project report by the person responsible for cost control. In addition to using the formal cost reporting systems, a wise cost controller also walks the places of work where key decisions are being made. By examining the actual construction work, drawings, programmes and estimates, and asking relevant questions everyone should become aware of their own cost plan targets and take them seriously.

When deviations from cost targets are identified (and this should be a rare occurrence provided that the targets are set in agreement with the individuals responsible for the work covered by the target), the following steps should be considered.

- If an estimate of the cost of a proposed decision exceeds the allowance in the cost plan, the decision should be reconsidered to find a more cost-effective alternative.
- If it proves impossible to contain a part of the project within its cost target, part of the contingency fund must be used.

- If an estimate falls significantly below the allowance in the cost plan, and upon careful checking everything has been taken into account, the saving should be added to the contingency fund.
- If it is impossible to contain overall costs within the overall budget, even by using the contingency fund, the client must be asked to increase the budget or accept a reduced project.

As a matter of formal routine, the designers should receive a cost report every month. It should contain no surprises. Ideally it will report that the project's costs are on target. If there are problems, the client should have been involved in the discussions to find an answer before receiving the formal report.

10.5 Control of change

In controlling change, three separate issues must be dealt with:

- changes or variations,
- development of the design,
- a request from the site for information.

Each must be clearly separated and their effect on the design process understood and assessed. They are all potentially disruptive to the design and construction processes. If the type of change is not clearly identified, confusion can follow, because each source has different cost consequences for the parties.

10.5.1 Changes or variations

Changes and variations are revisions to the design. These occur when the scope of the work changes or when the client revises the requirements for the building. Occasionally, legislative changes can have an impact on earlier design decisions, but transition arrangement usually minimize such impacts. Variations can occur if the briefing stage has been inadequate, or the client organization is not managed effectively enough to give consistent and timely instructions. Under these circumstances, the project manager must discuss the proposed changes with the client and agree on a course of action to be taken.

Alternatively, the client may reserve the right to vary the scope of the work at a late stage in the design to preserve flexibility, for example to meet changing letting conditions. However, as long as the design and construction team understand the flexibility that is required and the client accepts the resulting costs and uncertainty, a good working relationship can be maintained.

Good client communication to the whole of the design and construction team is essential to ensure that they appreciate the reasons for the client's uncertainty and late decisions. Otherwise, morale will suffer from the frustrations of attempting to achieve completion against unreasonable deadlines.

10.5.2 Design development

A difficult problem to manage is where shop and manufacturing information is used, either to develop aspects of the design team's information or to ensure that the manufactured products are compatible with one another. It is only when the fine detail necessary for making the product is available that the detailed co-ordination between components can take place. This in turn may require adjustments to be made to the products or to other parts of the design. Indeed, the designer may prefer to wait until the information from the fine detail drawings becomes available before even starting sections of the detail design, to avoid abortive work.

Either way it has to be recognized that a full set of production information is assembled from a wide range of different contributions and that the primary designers require large amounts of information to help in developing the whole set of information. No one, in these circumstances, can get it right the first time unless every aspect is planned meticulously.

A symptom of insufficient planning is the reissue of a large number of drawings containing continuous development during the construction stage. If this should happen, the design manager and the design team should review their planning process and adopt one of the methods described on page 135.

10.5.3 Requests from the site for information

With complex or intricate design it is very difficult to detail every aspect of the project on paper and there may be occasions where a detail may need to be resolved with the site team. A drawing may then be issued or a verbal instruction confirmed. The important point is that the decision should be made quickly to enable the work to continue. If a great deal of the design detail needs to be resolved in this way then the management of the design process has failed. A designer with executive decision-making responsibility should then be based on site or within easy reach.

10.5.4 Change control procedures

The need for a variation can originate from the client, architect, engineering consultants, specialists or contractor. The procedure will vary slightly in each case, but it must always be managed and the effect of any change should be understood and accepted by the client.

There is a fine line between design development and the clarification of intent and the design team must be aware of the distinction at all times. Also, any variation to a contract, once it has been signed, must be avoided unless the full consequences are understood and accepted by all parties before the change order is issued. Contracts must also recognize the inevitable flexibility needed and they must be managed accordingly. The primary need is to identify the source of the requirement for change and establish its significance both materially and contractually. The procedure shown (see Figure 32) formalizes an often *ad hoc* process.

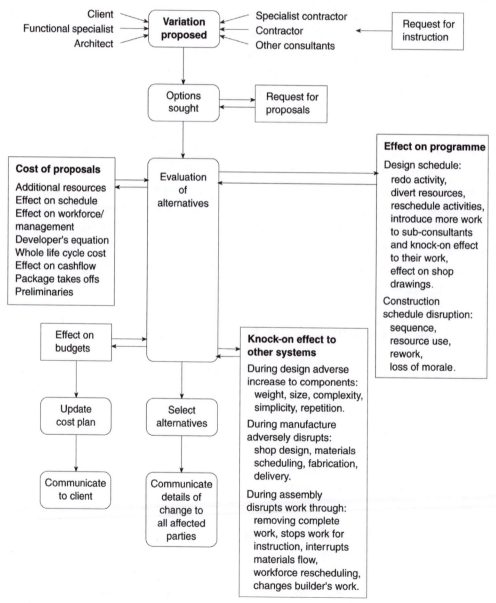

Fig. 32. Change order control flow chart.

Depending on the nature of the project and the problem, there may be a need to obtain input and information from specialists and a formal 'Request for Proposals' (RFP) may be issued. This is more common on managed forms of contract, but it is good practice on all types of contract and enables the change procedure to be initiated in a controlled way. The RFP states the scope of the proposed change and requests suggestions as to how it may be resolved and accommodated into the design.

The evaluation of the options is a complex matter because of the effect, not only on the design, but also on other designers and their work and, if they are already underway, the construction operations. The hidden effects of disrupting the supply and manufacturing processes must also be considered. The complete design team should evaluate the time and cost implication of each option. A formal evaluation and selection process should be adopted which leads to the decision whether or not to accept the change.

It is likely that small changes will be approved as and when they occur and that large changes will be considered more carefully. Care must be taken to monitor this strategy, as many minor changes have a more disruptive effect than a few major well-considered ones. It is essential to avoid insidious and continuous change. All changes must be identified and communicated to all those affected, in a clear and unambiguous way, acknowledging the implications on cost and progress.

10.5.5 Client functional specialists

The changes may be proposed by any of many organizations, but the functional specialists, i.e. in-house design groups or specialist engineering sections, within the client organization are very difficult to identify in terms of their authority over the project. The functional specialists are employed by the client as an integral part of its organization and thus have an implicit, though often ambiguous, power of decision on some or many aspects of the project.

While this is often to the benefit of the project it can become uncontrolled, and decisions made by functional specialists, either singly or in concert, can make the line of instruction from the client very confused indeed. This should be guarded against by carefully organizing the relationships within the client organization, as outlined on page 65.

10.6 Meetings and decision control

The formal communication system needs to be matched closely to the needs of the project. While informal communication is more powerful, it is sometimes dangerous in that it can become uncontrolled and rely too heavily on subjective judgement. The formal system must be able to respond quickly, or formal bypass routines may be needed for emergency action. Essentially there are two types of meeting where decisions are taken: co-ordination and control.

10.6.1 Design co-ordination meetings

At design co-ordination meetings the predominant objective is to co-ordinate the decision-making of the design team. The person chairing the meeting should be the most significant designer at that time. Preparations must be made for the meeting and only those who can contribute in a positive and constructive way should attend. Personal assistants to, and representatives of, participants are no substitute for decision-makers with responsibility. A degree of control will be placed on others, but the object is to arrive at a mutually agreed plan for the next series of design activities. The management role is to ensure that the project objectives are not compromised.

10.6.2 Design progress control meetings

At progress control meetings the review of progress compared with the programme must be held at the appropriate level and with the right people present, as the action to resolve failures can only be taken properly by those who have executive authority. For example, the formal system for monitoring and controlling the engineering information stage should be at the level of the zone or specialist package managers. The management role is to ensure that the people responsible for producing the design are actually doing it at the right time.

A positive approach to solving problems is always required. The manager has to guide others to the identification of the problem and what they should be doing to solve it. This is difficult. It always helps if those who have an intimate knowledge of the situation, and who can take responsible action, are present at the meeting. Time is then not wasted referring problems to those with the right authority.

10.6.3 Meeting practice

The basic rules are as follows.

- Every meeting should have a clearly stated purpose within either of the two categories discussed above.
- A purposeful agenda must be prepared beforehand.
- Time must not be wasted either through an ill-defined purpose or the introduction of extraneous matters, which are more suited to another kind of discussion.

- Determine the correct time for the meeting agenda and stick to it, as too much precious time is often wasted because of poor time keeping.
- The chairperson must have knowledge and control of the resources and be able to commit the designer or the trade contractor to the agreed action.

Another type of control meeting is the task force. Intensive meetings and data collection are used to resolve intractable problems or crises by bringing together all those that can contribute to a problem in a meaningful way. This method of working is largely informal but the results should be presented formally to avoid disrupting the existing working arrangements.

11

Design management
in action

The objective of this chapter is to place the issues that have been dealt with into a project process that incorporates the role of design management. The process map that has been developed primarily considers the design steps, but because they are inextricably interwoven with the whole project process there is an interface with other processes. The map is shown in Figure 33. The problem with this representation is that it is simplistic, because to map the whole process in its entire detail would be too complex to grasp at the strategic level. However, it must be recognized that while this process map is one version, and fairly robust, it must be tailored to the specific case and project needs.

When developing a specific project plan, it is important to identify key points at which the process is drawn to a conclusion and everyone involved signs-off the stage. By achieving an explicit commitment to the decisions so far, the management process of the subsequent stages is eased, because at least there are some points of control in an otherwise fluid process. The importance of this cannot be over-emphasized. In the instance where sign-off is nearly agreed (some use the phrase, 'soft sign-off'), but not actually agreed, the result is that uncertainty continues and disrupts the subsequent processes. A total map will complicate the process further because at certain points, for example the consultant's detail design, the package supply chains will emerge and fragment and each chain will progress through the subsequent stages at different rates. This will be further compounded as individual components emerge and their process will also go through the downstream stages at their own rate.

It will probably be quite rare for the project process to start off in such an orderly way that everyone is involved fully from the start. It would be sensible for each organization, as it is involved in the process, to perform an audit of what has happened already and to determine where there are gaps in the quality of the information. The gaps need to be filled before the organization commits itself to the project.

A feature of this process map is that it considers the changing relationships between the main parties at each of the major change points in the process. Just to reiterate the discussion in Chapter 4, the three roles of client (C), designers (D), and management (M)

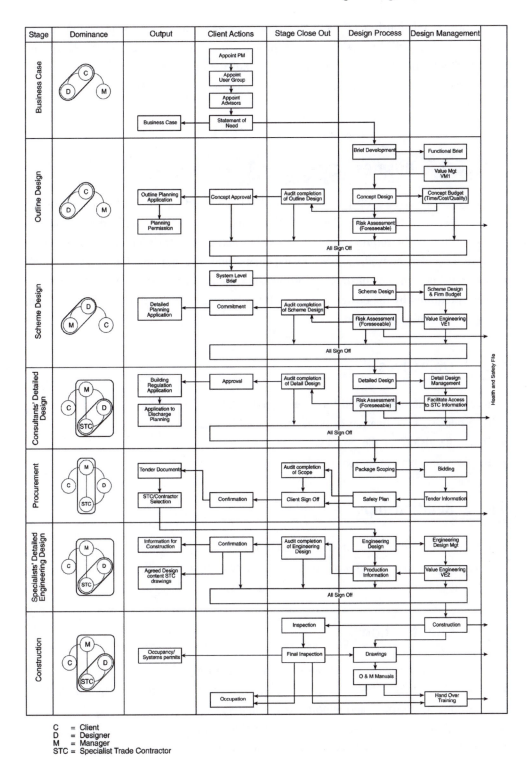

C   = Client
D   = Designer
M   = Manager
STC = Specialist Trade Contractor

Fig. 33. A design management process map (see also inside back cover).

are shown in relationship to each other, but change in terms of their prominence in decision-making and leadership as the project proceeds. An additional role of the specialist trade contractor (STC) is included to indicate the increasing importance of their specialist contribution to the design process. Each of the main stages is considered in detail below.

11.1 The business case

This stage is involved with placing the need for the building within the context of the business and the user's needs. Clients need to consider how their organization is going to be influenced by the requirements of the design and construction process. Equally they need to consider how their own management and decision-making processes will influence the design and construction process. If both do not recognize each other then the decisions required by the design process will not be made, either at the right time or with the right level of commitment.

11.1.1 Objectives

- To establish an agreed need for a new build solution.
- To produce a specification of need and agreed business case.
- To set the vision for the project.
- To produce a plan of action to achieve the business case.

11.1.2 Relationships

- The client organization is in the dominant role.
- The designers and managers may be required to assist with the business case development.

11.1.3 Client action

- Determine the scale of input from their own organization.
- Appoint a client's representative or project manager with executive responsibility.
- Determine user groups and appoint representatives.
- Clarify decision-making structure and respective authority.
- Identify key decision-makers and involve appropriately.
- Confirm the statement of need.
- Agree the business case and the need to proceed.

11.1.4 Design team action

- Audit state of client's decision-making progress.
- Assist client groups in determining needs.
- Help assess the validity of the client's needs.
- Provide alternate solution strategies.
- Help identify high value returns and options.

11.1.5 Design management

- Implement formal stage start-up meeting/away days to achieve team integration.
- Establish programme for decision-making.
- Identify specialist design and skills requirements and help obtain them.
- Help identify constraints, both business and physical.
- Identify performance benchmarks for the design and construction process.
- Assist in the preparation of the statement of need.

11.1.6 Stage close out

- A validated business case and statement of need signed-off by the key decision-makers within the client organization.

11.2 Outline design

The business case sets the strategy to achieve the need. This must now be turned into a detailed evaluation of the specific needs of all of the users. The initial brief must be developed. In most business situations this may well change as the project develops over time. Unless the project is one of a series it is unlikely that every aspect of the client's requirements will be identified at this point. Therefore a decision must be made as to the level of detail to which the brief is to be developed at this stage. It would, of course, simplify the management task by increasing the certainty if every aspect could be decided in the one brief at the outset. This is unlikely, so a staged brief process is more likely.

The involvement of the design team at this stage will be crucial to the success of the project. Their recruitment is a lengthy process and the client organization needs to be aware of the potential impact that a successful working relationship could have on the outcome of the project. The project manager must put in place the criteria for selection so that the client's key executives can start the process of discussion and selection. A major criterion must be the development of a mutual understanding of design, the designer's goals and capabilities, together with the client's goals and constraints. Once a number of designers sympathetic to the client's ambitions are identified a shortlist for final selection can be prepared.

The business case will contain the essential value statements for the project. This stage of the process, via value management workshops, helps to set the criteria for assessing the value of the various statements and enables the priorities to be set.

11.2.1 Objectives

- To establish the detailed brief and functional brief.
- Set priorities for satisfaction in the brief.
- To develop the concepts and characteristics of the proposed building.
- To submit and obtain outline planning approval.

11.2.2 Relationships

- The client organization is in the dominant role.
- The designers and managers will be totally involved with the client in the development of the brief and the assessment of feasibility against the business case.

11.2.3 Client action

- Set criteria for the design team selection.
- Establish a methodology for developing mutual design objectives with the designers.
- Select the lead designers for the key building systems.
- Set up user groups to define the functional needs/spaces and relationships.
- Establish a working group to assess the impact of the new building on the current working of the business.
- Establish priorities within the value statements for the project.
- Prioritize the functional specification.
- Seek to use the best practices of the construction industry.
- Appoint a Planning Supervisor.

11.2.4 Design team action

- Implement a formal stage start-up meeting/away days to achieve team integration.
- Develop mutual design objectives with the client.
- Contribute to the preparation of the brief.

- Organize the production of sketch designs, specification and schedules.
- Obtain geological information.
- Determine the requirements of statutory authorities, planning and other bodies.
- Determine the overall space planning of the building and derive net/gross ratios.
- Provide designs for planning applications and negotiations with other bodies.
- Complete the risk assessment of foreseeable construction health and safety risks.

11.2.5 Design management

- Implement a formal stage start-up meeting/away days to achieve team integration.
- Determine the information needs and sources and advise on their provision.
- Implement co-ordinated data management and drawing protocols for all project participants.
- Produce a strategic plan of the processes of: design, procurement and construction.
- Provide the meeting and decision-making structure.
- Assist/facilitate the value management workshops.
- Fully scope and document risks.

11.2.6 Stage close out

- Audit developed concept design against priorities in the brief.
- Outline planning approval achieved.
- Concept and brief signed off by all parties.
- Challenging targets set to obtain leading edge practice from the design and construction process.

11.3 Scheme design

At the scheme design stage, all the major systems in the building are specified and integrated into the final project scheme. It is an intensive stage of evaluation of alternative technological solutions.

11.3.1 Objectives

- To develop the systems level briefs if necessary.
- To develop the elemental cost plan.
- To value engineer the project within the challenging targets.
- To submit and obtain detailed planning approval.

11.3.2 Relationships

- Designers lead this stage in conjunction with the managers who provide process management and also production method evaluation.
- The client is closely involved in decision-making and is kept informed.

11.3.3 Client action

- To approve system level briefs and designs.

11.3.4 Design team action

- Develop the design to finalize the building's appearance, layout, choice of technological systems, and materials.
- Ensure the design meets the constraints within the cost plan.
- Identify specialist knowledge and information needs.
- Negotiate with the planning authority to achieve detailed planning approval.

11.3.5 Design management

- Implement formal stage start-up meeting/away days to achieve team integration.
- Identify principal supply chains and key interface points between them and the design process.
- Develop design information production schedule which recognizes the supply chain needs.
- Define deliverables of each designer at each interface.
- Chair/attend design team progress meetings.
- Chair/attend design team working meetings.
- Provide/facilitate value engineering workshops.

11.3.6 Stage close out

- Audit developed scheme design against priorities in the brief.
- Detailed planning approval achieved.
- Scheme signed-off by all parties.

11.4 Consultant's detailed design

Once planning has been approved, the issue is to develop the detail of the systems in the building to ensure that they are fully co-ordinated, tolerances are compatible and any minor adjustments of spaces are made.

11.4.1 Objectives

- To obtain building regulation approvals.
- To obtain rationalization and integration of sub-systems.
- To provide a health and safety plan.
- To provide scoping information to appoint specialist trade contractors

11.4.2 Relationships

- The management processes become dominant.
- Specialist trade contractors are involved with the designers in providing advice and knowledge of their systems.
- The client is involved at key decision points.

11.4.3 Client action

- To chair design review meetings.

11.4.4 Design team action

- Issue request for proposals to specialist trade contractors.
- Ensure all systems are fully co-ordinated at the detail level.
- Submit requests for modification to the client for approval.
- Assessment of foreseeable risks.

11.4.5 Design management

- Develop and monitor design information supply plan.
- Audit the design to achieve production targets.

11.4.6 Stage close out

- Provide information to meet the needs of the procurement process.

11.5 Procurement

A crucial stage in the effective development of the design is the purchasing of the technical knowledge input from specialist manufacturers and trade contractors. Procurement may need to occur in several phases rather than at one single point because the needs of the design process and supply chain interact. The design process requires a large amount of technical input. To expect this as a free service is unrealistic and at best only limited information will be provided. Further iterations of the design, as the information becomes more available once the supplier is contracted, will be necessary unless the supplier is involved correctly.

Procurement of technical information will inevitably become entwined in the procurement approach for the whole project. Where individual package procurement is not being used it is essential to ensure that the procurement and contracting approach takes account of the design process, and this can only be done if there is a comprehensive plan of the design process.

11.5.1 Objectives

- To identify the knowledge inputs for advanced procurement.
- To sequence procurement based on design interface needs and long delivery scheduled items.

11.5.2 Relationships

- The management processes dominate.
- The designers drive the information need that structures the procurement process.
- The client approves the procurement process and structures its decision-making accordingly.

11.5.3 Client action

- To agree the procurement strategy and sign-off the package at the appropriate time.

11.5.4 Design team action

- Identify scope of packages.
- Provide all information necessary for each package.
- Specify the information needs from each package supplier.

11.5.5 Design management

- Identify scope of packages.
- Resolve design liability issues.
- Identify and specify the information needs from each package supplier together with the information needs of each package supplier.
- Identify contractors, suppliers and specialists in respect of their capability to meet all project needs.
- Shortlist suitable contractors, specialists and trade contractors.
- Assist in the development of the bidding process.
- Identify the schedule for information supply from each package that respects interface issues both between package and system lead designer and the interfaces between packages.

11.5.6 Stage close out

- Technical design information is identified and bought in relation to information needs and quality of input.

11.6 Specialist's detailed design

As buildings grow in complexity, the input of the specialists to the resolution of the working details of the components and their interaction becomes more important. The process involves several stages each with its own requirements. In the UK over half of the total project information is produced by the specialists, while in the USA it can be over 80%. However, the two systems are not the same. In the USA the design team produces relatively few drawings and the specialists provide nearly all of the design, but only through known and acceptable 'standard' solutions. In the UK, projects use much higher levels of innovative application of 'standard' technologies, thus the relationship with the specialists is very different.

The interaction with the specialist takes place through three stages.

- Discussion about the feasibility of using their products and techniques.
- The development of the individual components, and ensuring fit and fixing co-ordination.
- The development of manufacturing or shop drawing information to enable the individual components to be manufactured.

An efficient design process recognizes the implication of these three stages and ensures that there is a coherent and integrated chain of responsibility that ensures that the initial values and decisions are not compromised as the process moves from outline to detail.

11.6.1 Objectives

- To develop the design to enable the manufacture of every component in a defect-free manner to the schedule required so that delivery meets the site assembly schedule.

11.6.2 Relationships

- The management processes dominate to ensure that the specialist trade contractors and the designers work together to achieve the project objectives.
- The client is closely involved in preserving the priorities set in the brief, cost plan, and authorizing trades between them if necessary.

11.6.3 Client action

- To make decisions as required by the development of the detail design.
- To control the need of the business to modify or change the detailed needs.

11.6.4 Design team action

- Work with specialists to maximize their capability, manufacturing processes and production systems.
- Work within a formal proposal and approval system to ensure design development is carried out within the agreed package or elemental budget.
- Co-ordinate all design information to ensure completeness, accuracy and interface compatibility.

11.6.5 Design management

- Implement formal stage start-up meeting to achieve team integration of system level groups.
- Integrate suppliers' and specialists' design and production programme.

- Implement document control system for the issue of information between the design team and specialists.
- Supervise the Requests For Proposals (RFP) to ensure that all requests are evaluated within the context of the brief and elemental cost plan, and that they are approved before implementation.
- Supervise the Requests For Information (RFI) from the specialists to ensure that all requests are evaluated within the context of the brief and elemental cost plan, and that they are approved before implementation.
- Ensure that all project managers are kept informed of all discussions that affect their package.

11.6.6 Stage close out

- Package and production information is signed-off for manufacture or assembly.

11.7 Construction

In practice over 70% of the total information for the project will be generated and issued once construction has commenced. All of the major decisions will have been made, but there will be a myriad of detailed issues to resolve. Most of the issues will be at the level of the component and subcomponent interface with a constant need for the designers, and possibly the client, to be available to make rapid decisions.

11.7.1 Objectives

- To resolve all Requests For Information (RFI) as required by the demands of the manufacture and assembly process.

11.7.2 Relationships

- The management processes will be working closely with the specialists to ensure that the approvals that they need are obtained in sufficient time to achieve an uninterrupted flow of materials, components and assemblies.

11.7.3 Client action

- To make decisions as required at the speed needed by the assembly process.
- To postpone all major changes until after the completion of the main project or divide the fit-out stage into a separate sub-project.

11.7.4 Design team action

- Provide a team on or near the site to ensure decisions are made as close to the point of need as possible.
- Devolve authority for making decisions about site assembly to the lowest point possible in the design hierarchy.
- Produce clear testing and commissioning routines.
- Complete as-built drawings.
- Develop operation and maintenance manuals.

11.7.5 Design management

- Implement a document control system for the issue of information for construction.
- Determine roles and responsibilities for the testing and signed-off approvals of the building and its systems.
- Provide the Planning Supervisor with all certificates, as-built drawings, and operating manuals.
- Complete a risk assessment file and pass to the Planning Supervisor.
- Train users in the operation of the building and its systems.

11.7.6 Stage close out

- Building handed over with zero defects

11.8 Final comments

The management of the design process is a subtle and complex problem. Although the historical context of construction design is one in which designers were also managers of the process, the complexity of the task has grown to a point where most people recognize the need for management. However, design needs a very careful kind of management. While acknowledging the enormous difficulties associated with generalizing the process to the extent that we have done here, we hope that the advice in this book will provide a starting point in helping project teams to continue to provide clients and society with wonderful, sensitive and useful building designs.

Glossary

This glossary is presented in two sections. First, are terms relating specifically to the management of building design and second, are terms relating to management and organizational theories.

Building design management terms

Architect, designer and design professionals
The architect or other professional design consultant employed to design the building by leading a design team comprising design consultants and, where necessary, specialist contractors.

Client
The individual or organization commissioning the building project and directly employing the designer, the project manager and the works contractors.

Concept architect
The architect who develops the vision of the design to meet the client's requirements.

Construction manager
The individual or organization employed by the client to enhance their ability to control the cost and to manage the contract in co-operation with the designer, through the works contractors.

Cost consultant
The individual or organization that may be appointed by the client to provide cost advice.

Design manager
The manager responsible for co-ordinating the design task to ensure that information of the appropriate quality is delivered within the project time-scale to meet the needs of the design, manufacturing and construction process.

Design team	The group of individuals drawn from contributory professional practices who will work together to provide the concept, scheme and detailed design information.
Design team leader	The individual, probably the architect, who will co-ordinate and manage the design team.
Engineering	The development of the scheme design into the production information for manufacturing and construction which considers the integrated relationship between user needs, quality, value, and construction practicality and efficiency.
Planning supervisor	This role is defined in UK Health and Safety law, as represented by the Construction (Design and Management) Regulations. This person ensures that the way the building is designed takes account of the health and safety of all those involved in the construction process, maintenance, use and demolition.
Principal contractor	This role is defined in UK Health and Safety law, as represented by the Construction (Design and Management) Regulations. This person ensures that the way the building is built takes account of the health and safety of all those involved in the construction process, maintenance, use and demolition.
Project director	The person acting on behalf of the project client who provides the overall direction for converting the client's expectations into an achievable output.
Project manager	A manager within any organization who has total responsibility for delivering the project or contribution to the project, within the time, cost, and quality parameters of the project.
Specialist trade contractor	A works contractor with special design knowledge that may be called upon by the designer to assist in pre-construction design.
Trade contractor	A works contractor supplying products or trades to a given specification.
Works contractor	The individual or commercial organization employed to complete the working design to meet the designer's specification and to produce a site-assembled work package or temporary works.

Management and organizational terms

Contingency theory
The idea that the way an organization is structured is primarily dependent upon the operating environment of the organization, the key variable of which is usually seen as technology.

Co-ordination
A management role of ensuring that output from each participant is oriented towards the organization's objectives.

Differentiation
Differences in the technology, attitudes, allegiances and behaviour of people involved in organizational processes.

Environment
That which is beyond the boundary of a system.

Hierarchy
The positioning of managers over staff so that the diverse inputs of subordinates can be brought together.

Integration
The management role that brings diverse contributions together into a cohesive whole, a response to differentiation.

Manage
To conduct things and people in order to achieve some end.

Multi-organization
A collection of firms and organizations in a structured relationship that is brought together for the purpose of designing and constructing.

Organization
A collection of individuals in a structured relationship that is brought together in order to achieve some aim.

Organizational structure
The pattern of the relationship between people and tasks within an organization.

Organize
To arrange related parts to form a co-ordinated whole.

Planning
Scheduling and resource planning of work to be done.

Procedures
Standard ways of doing things

Satisficing
The process of choosing from among a variety of conflicting needs a solution that is not the best answer to any particular need, but offers the least compromises overall.

Systems theory
The body of theory that analyses things by looking at them as a group of inter-related parts within a boundary.

References

Addis, W. (1996) *The Art of the Structural Engineer*, Artemis Limited, London.

Akin, O. (1986) *Psychology and architectural design*, Pion, London.

Dowson, P. (1990) *Design Delegation*, 192, No 25 & 26, MBC Architectural Press and Building Publications, London.

Austin, S., Baldwin, A., Li, B. and Waskett, P. (1999a) Analytic design planning technique for programming building design. *Proceedings of the Institute of Civil Engineers, Structures & Buildings*, **134**, pp. 111–18.

Austin, S., Baldwin, A., Li, B. and Waskett, P. (1999b) Analytic design planning technique (ADePT): a dependency structure matrix tool to schedule the building design process. *Construction Management and Economics*, **18**(2), 173–82.

Austin, S., Baldwin, A., Li, B. and Waskett, P. (1999c) Analytic design planning technique: a model of the detailed building design process. *Design Studies*, **20**, 279–96.

Banwell, H. (1964) *The Placing and Management of Contracts for Building and Civil Engineering*, London; HMSO.

Benhaime, M. (1997) *Inter-firm relationships within the construction industry: towards the emergence of networks? A comparative study between France and the UK*, DBA thesis, Brunel University.

Bennett, J. (1992) *International Construction Project Management*, Butterworth-Heinemann, London.

Bennett, J. and Jayes, S. (1998) *The Seven Pillars of Partnering*, Thomas Telford, London.

Binney, M. (1991) *Palace on the River*, Wordsearch Publishing, London.

Bleeke, J. and Ernst, D. (1995) Is your strategic alliance really a sale? *Harvard Business Review*, **73**(1), 97–105.

Broadbent, G. (1973) *Design in Architecture*, Wiley, Chichester.

Chapman, R. (1997) Managing Design Change. *Architect's Journal*, June, 47–9.

Coase, R. H. (1937) The nature of the firm. *Economica*, **4**, 386–405.

Connaughton, J. N. and Green, S. D. (1996) *Value Management in Construction: a Client's Guide*, Construction Industry Research and Information Association, London.

Construction Industry Research and Information Association. (1983) *Client's Guide to Management Contracts in Building*. Special Publication No 33, London.

Construction Industry Board. (1996) *Selecting Consultants for the Team: Balancing Quality and Price*, Thomas Telford, London.

Construction Industry Board. (1997) *Briefing the Team*, Thomas Telford, London.

Cornick, T. (1990) *Quality Management for Building Design*, Butterworths Heinemann Ltd, Guildford.

Cross, N. (1996) Creativity in Design: not leaping but bridging, Proc: Creativity and Cognition, (Candy, L. and Edmonds, E., eds) pp. 27–35.

Cyert, R. M. and March, J. E. (1963) *A behavioural theory of the firm*, Prentice-Hall, Englewood Cliffs, NJ.

Darke, J. (1984) *Developments in Design Methodology*, Wiley, Chichester.

Dawson, S. (1996) *Analysing Organizations*, 3rd edn, Macmillan, Basingstoke.

Duffy *et al.*, (1995) *The Review of the Profession, Stages Two and Three*, RIBA, London.

Dumas, A. and Mintzberg, H. (1992) Managing the form, function and fit of design. *Design Management Journal*, **2**(3), 46–54.

Egan, J. (1998) *Rethinking construction*, Dept. of Environment, Transport and the Regions, London.

Flanagan, *et al.*, (1979) *UK and US construction industries: a comparison of design and contract procedures*, Royal Institution of Chartered Surveyors, London.

Flanagan, R. and Norman, G. (1983) *Life Cycle Costing for Construction*, RICS, London.

Fowler, C. (1997) *Building Services Engineering, a Strategy for the Future*, Reading Production Engineering Group, University of Reading.

Frampton, K. (1994) *Modern Architecture – a Critical History*, Thames and Hudson Ltd, London, p. 161.

Galbraith, J. (1973) *Designing Complex Organizations*, Addison-Wesley, Massachusetts.

Gidado, K. I. (1996) Project complexity: the focal point of construction planning. *Construction Management and Economics*, **14**(3), 213–25.

Gray, C. (1996) *Value For Money – Helping the UK Afford the Buildings it Likes*, Reading Construction Forum, The University of Reading.

Gray, C. (1987) *Buildability – the Construction Contribution*, Occasional paper no. 29, CIOB, Ascot.

Gray, C., Hughes, W. and Bennett, J. (1994) *The successful management of design*, Centre for Strategic Studies in Construction, Reading.

Green, S. D. and Popper, P. A. (1990) *Value Engineering – the Search for Unnecessary Cost*, Occasional Paper no. 39, CIOB, Ascot.

Gutman, R. (1988) *Architectural Practice*, Princeton Architectural Press, New York.

Handy, C. (1986) *Understanding Organizations*, Penguin Business, St. Ives.

Handy, C. (1990) *The Age of Unreason*, Arrow, London.

Hardington, S. (1995) *A Guide to Recent Architecture – England*, Ellipses London Ltd, London.

Harvey-Jones, J. (1988) *Making it Happen: Reflections on Leadership*, Fontana, London.

Hawk, D. (1996) Relations between architecture and management. *Journal of Architectural and Planning Research*, **13**(1), 10–33.

Hickling, A. (1982) Beyond a linear iterative process. In: Evans *et al.* eds, *Changing Design*, Wiley, Chichester.

Hollowman, C. and Hendrick, H. (1971) Problem solving in different size groups. *Personal Psychology*, **24**, 489–500.

Honeyman, S. (1990) *Construction Management Forum Report and Guidance*, Centre for Stategic Studies in Construction, Reading.

Hughes, W. P. (1989) Organizational analysis of building projects. PhD thesis, Department of Surveying, Liverpool Polytechnic.

Hutton, W. (1996) *The State we're in*, Vintage, London.

Jaques, E. (1976) *A general theory of bureaucracy*, Gower, Aldershot.

Joint Contracts Tribunal (1998) *Standard form of building contract: with quantities*. Joint Contracts Tribunal, London.

Latham, M. (1994) *Constructing the Team* (Final report of the government/industry review of procurement and contractual arrangements in the UK construction industry), HMSO, London.

Lawrence, P. R. and Lorsch, J. W. (1967a) New management job: the integrator. *Harvard Business Review*, **45**, 142–51.

Lawrence, P. R. and Lorsch, J. W. (1967b) *Organization and Environment: Managing Differentiation and Integration*, Harvard University Press, Massachusetts.

Lawson, B. (1994), *Design in Mind*, Butterworth Architecture, Oxford.

Lawson, B. (1980) *How Designers Think*, Butterworth Architecture, London.

March, L. (1976) The logic of design and the question of value. In: March, L. (ed.), *The Architecture of Form*, Cambridge University Press, Cambridge, UK. [Reprinted in Cross, N. (ed.) (1984) *Developments in Design Methodology*, John Wiley and Sons Ltd, Chichester, UK.]

Martin, V. and Ishii, K. (1997) *Design for Variety: the Development of Complexity Indices and Design Charts*, in: Procs. 1996 ASME design engineering technical conferences and computers in engineering conference August 18–22, Sacramento, CA.

Mills, D. Q. (1991) *Rebirth of the Corporation*, John Wiley and Sons Inc. In Peters p. 245.

Mintzberg, H. (1991) Effective organization: forces and forms. *Sloan Management Review*.

Morgan, G. (1986) *Images of Organizations*, Sagel, London.

Moxley, R. (1984) *The Architect's Guide to Fee Negotiations*, Architectural & Building Practice Guides, London.

Murdoch, J. and Hughes, W. (2000) *Construction contracts: law and management*, 3rd edn, Spon, London.

Ove Arup & Partners (1985) *Technology in Architecture*, Ove Arup & Partners, London.

Pascale, R. (1990) *Managing on the Edge: How Successful Companies use Conflict to Stay Ahead*, Penguin, Harmondsworth.

Peters, T. (1992) *Liberation Management – the Necessary Disorganization for the Nanosecond Nineties*, Macmillan Press, London.

Pevsner, N. (1968) Studies in art, architecture and design, Vol. 2: *Victorian and after*, Thames and Hudson, London.

Pine, B. J. (1993) *Mass Customisation: the New Frontiers in Business Competition*, Harvard Business School Press, Boston, MA.

Port, S. (1989) *The Management of CAD*, BSP Professional Books, Oxford.

Farrell, R. (ed.) (1968) *Power Station Construction* CEGB, London.

RIBA (1991) *The Architect's Handbook of Practice and Management*, RIBA Publications Ltd, London.

Risbero, B. (1982) *Modern Architectural Design – an Alternative History*, The Herbert Press, London, p. 180.

Rougvie, A. (1981) *Project Evaluation and Development*, Mitchells, London.

Schön, D. A. (1983) *The Reflective Practitioner*, Temple Smith, London.

Schrage, M. (1990) *No More Teams: Mastering the Dynamics of Creative Collaboration*, Doubleday, New York.

Shand, P. (1954) *Building: the Evolution of an Industry*, Token Construction, London.

Sonnerwald, D. H. (1996) Communication roles that support collaboration during the design process. *Design Studies*, **17**, 277–301.

Southwell, M. (1997) *Projects and the Management of Complexity*, MSc thesis, Department of Construction Management & Engineering, University of Reading.

Staveley, H. S. (1991) *Access to Buildings for Inspection and Maintenance*, Technical Information Service Number 136, CIOB, Ascot.

Thompson, R. D. (1967) *Organizations in Action*, McGraw-Hill, New York.

Tjosvold, D. (1985) Implications of controversy research for management. *Journal of Management*, **11**, 21–37.

Trinh, T. T. P. and Sharif, N. (1996) Assessing construction technology by integrating constructed product and construction process complexities: a case study of embankment dams in Thailand.

Construction Management and Economics, **14**(6), 467–84.

Williams, S. (1989) *Hong Kong Bank – the Building of Norman Foster's Masterpiece*, Jonathan Cape, London, p. 61.

Williamson, O. E. (1975) *Markets and Hierarchies – analysis and anti-trust implications: a study in the economics of internal organization*, Free Press, New York.

Womack, J. P. and Jones, D. T. (1996) *Lean Thinking – banish waste and create wealth in your corporation*, Simon & Shuster, London.

Woodward, J. (1965) *Industrial Organization: Theory and Practice*, Oxford University Press, London.

Index